Urban Utopias

Utopia tends to generate a bad press – regarded as impracticable, perhaps nostalgic, or contradictory when visions of a perfect world cannot accommodate the change that is necessary to a free and self-organizing society. But people from diverse backgrounds are currently building a new society within the old, balancing literal and metaphorical utopianism, and demonstrating plural possibilities for alternative futures and types of settlement. Thousands of such places exist around the world, including intentional communities, eco-villages, permaculture plots, religious and secular retreats, cohousing projects, self-build schemes, projects for low-impact housing, and activist squats in both urban and rural sites. This experience suggests, however, that when planning and design are not integral to alternative social formations, the modern dream to engineer a new society cannot be realized.

The book is structured in four parts. In Part One, literary and theoretical Utopias from the early modern period to the nineteenth century are reconsidered. Part Two investigates twentieth-century urban utopianism and contemporary alternative settlements, focusing on social and environmental issues, activism and eco-village living. Part Three looks to wider horizons in recent practices in the non-affluent world, and Part Four reviews a range of cases from the author's visits to specific sites. This is followed by a short conclusion in which a discussion of key issues is resumed.

This book brings together insights from literary, theoretical and practical Utopias, drawing out the characteristics of groups and places that are part of a new society. It links today's utopian experiments to historical and literary Utopias, and to theoretical problems in utopian thought.

Malcolm Miles is Professor of Cultural Theory at the University of Plymouth, UK, where he convenes the Critical Spaces Research Group and coordinates a research methods programme for the Faculty of Arts.

Urban Utopias
The built and social architectures
of alternative settlements

Malcolm Miles

LONDON AND NEW YORK

First published 2008 by Routledge
2 Park Square, Milton Park, Abingdon, Oxon OX14 4RN

Simultaneously published in the USA and Canada
by Routledge
270 Madison Avenue, New York, NY 10016

Routledge is an imprint of the Taylor & Francis Group, an informa business

© 2008 Malcolm Miles

Typeset in Times New Roman by
Keystroke, 28 High Street, Tettenhall, Wolverhampton
Printed and bound in Great Britain by
Antony Rowe Ltd, Chippenham, Wiltshire

All rights reserved. No part of this book may be reprinted or reproduced or utilised in any form or by any electronic, mechanical, or other means, now known or hereafter invented, including photocopying and recording, or in any information storage or retrieval system, without permission in writing from the publishers.

British Library Cataloguing in Publication Data
A catalogue record for this book is available from the British Library

Library of Congress Cataloging in Publication Data
Miles, Malcolm.
Urban utopias: the built and social architectures of alternative settlements/Malcolm Miles.
p. cm.
Includes bibliographical references and index.
1. Utopias. 2. Utopias—Case studies. 3. Communal living.
4. Communal living—Case studies. 5. Collective settlements.
6. Collective settlements—Case studies. I. Title.
HX806.M45 2007
307.77—dc22   2007020418

ISBN10: 0–415–37575–4 (hbk)
ISBN10: 0–415–37576–2 (pbk)
ISBN10: 0–203–09912–5 (ebk)

ISBN13: 978–0–415–37575–7 (hbk)
ISBN13: 978–0–415–37576–4 (pbk)
ISBN13: 978–0–203–09912–4 (ebk)

Contents

List of illustrations   vii
Acknowledgments   ix

Introduction   1

PART ONE
Histories and theories   5

1   Imagining places: literary Utopias and the far-away   7
2   Drawing lines: modernity and Utopia   22
3   Planning harmony: Charles Fourier and utopian socialism   37

PART TWO
Practices   57

4   New cities   59
5   Social Utopias   75
6   Ecotopias: frameworks   91
7   Ecotopias: practices   110

PART THREE
Horizons   131

8   Mud-brick Utopias   133
9   A barefoot society   157

PART FOUR
Short case studies 177

Case 1	Economy, Pennsylvania, USA	179
Case 2	Arcosanti, Arizona, USA	184
Case 3	Auroville, Tamil Nadu, India	189
Case 4	Christiania, Copenhagen, Denmark	195
Case 5	Ufa-Fabrik, Berlin, Germany	201
Case 6	Uzupio, Vilnius, Lithuania	205
Case 7	Cambridge Cohousing, Massachusetts, USA	209
Case 8	Ecovillage at Ithaca, New York State, USA	214
Case 9	ZEGG, Belzig, Germany	220
	Conclusion	227
	Bibliography	233
	Index	248

Illustrations

Front cover The Visitor Centre, Auroville, Tamil Nadu, India
1 Charterville Allotments, Minster Lovell, one of the cottages 53
2 Milton Keynes, the oak tree in the mall 65
3 New Gourna, the mosque courtyard 141
4 New Gourna, street scene 144
5 New Gourna, recent house extension, exterior 145
6 New Baris, the agricultural cooperative 147
7 New Baris, villa, detail of roof terrace 149
8 SWRC, the old Fever Hospital 158
9 SWRC, street with post office 160
10 SWRC, solar cooker workshop 166
11 SWRC, geodesic domes and administrative building 168
12 SWRC, children at a night school 170
13 Economy, the community kitchen 179
14 Economy, the village pump 182
15 Arcosanti, the city on a hill 184
16 Arcosanti, the Modernism of cypress trees and concrete 187
17 Auroville, terrace of the Visitor Centre 190
18 Auroville, a house in improved vernacular style 193
19 Christiania, self-build, low-energy houses by the river 196
20 Christiania, preparation for a parade 200
21 Ufa-Fabrik, the organic café and bakery 201
22 Ufa-Fabrik, water treatment pools 203
23 Uzupio, road sign at the bridge 205
24 Uzupio, old houses by the river 208
25 Cambridge Cohousing, town houses 209
26 Cambridge Cohousing, the garden terace 213
27 Ecovillage at Ithaca, the common house seen from the second phase 214
28 Ecovillage at Ithaca, a street in the first phase 218
29 ZEGG, pile of woodchips and the converted boiler house 221
30 ZEGG, the treehouse for lovers 225
31 ZEGG, 'Another World is Possible' 229

32 ZEGG, members of the Global Ecovillage Network at the 2004 annual meeting 230
33 'Mapping the Future', Dave Beech (Freee Art Collective), Futurology project, Walsall Art Gallery, 2004 (photo A. Hewitt by permission of Freee) 232

Acknowledgements

I am grateful to the Arts and Humanities Research Council for a grant which enabled me to travel to alternative settlements in India, Europe and North America, and to see Hassan Fathy's mud-brick architecture in Egypt. These visits were invaluable in adding a practical dimension to the book's historical and theoretical reflections. I am especially grateful to all those I met who live in the alternative settlements visited – the Social Work Research Centre (Barefoot College), Tilonia, Rajasthan; Auroville, Tamil Nadu; Christiania, Copenhagen; ZEGG, Belzig, Germany; the ecovillage at Ithaca, New York State; Cambridge Cohousing, Massachusetts; Arcosanti, Arizona; and the Peace House, Coventry; and also to those whom I met at the 2005 annual meeting of the Global Ecovillage Network and from other alternative settlements such as Tinker's Bubble in Somerset and Findhorn in Scotland. Without the personal explanations and accounts of their principles and daily lives I would have understood little of the alternative society that quietly grows within today's mainstream society. I have been fortunate, too, to be able to rehearse some of the ideas on which the book is based in seminars in the Faculty of Arts at the University of Plymouth. I remain indebted to my academic colleagues – staff and research students – for encouraging me to develop a utopian strand in my work, in particular to Sarah Bennett, Robert Brown, David Coslett, Phillip Hawkins, Nicola Kirkham, Katy MacLeod and Robertas Mock.

Introduction

Background

Utopia tends to generate a bad press: it is regarded as impracticable, nostalgic, or contradictory; most often it is perceived as being unable (as a supposedly perfect world) to admit the change necessary in a self-determining society. Literary Utopias, such as Thomas More's *Utopia*, specify the daily life of the envisioned ideal society to an extent that is over-prescriptive and distant from contemporary reality. This level of prescriptive detail also tends to inform nineteenth-century utopian theory, in which designs for an ideal society obscure its inherent contradictions. The same desire to create a better world is central to Western Modernism, though in that case it is contradicted by a privileging of design and professional expertise over the tacit, experiential knowledge of those who live in alternative settlements. But literary accounts need not be taken literally – they may not be perfect worlds at all – and can be read in a more ordinary way as critical reflections on the writer's world. Similarly, utopian designs reflect perceived injustices in the conditions in which they were produced. And when utopian ideas idealize the past or an imagined future in aesthetic terms this, too, can be regarded as an oblique critique. As objects of study, all such utopian texts and images are open to critical understanding, though this may appear to achieve little in the face of an onslaught of social and environmental destructiveness. Hope may appear somewhat academic, or restricted to global activism and resistance to a system of which the default position is social division and environmental destruction. Yet there are thousands of alternative settlements today where people from diverse backgrounds have built a new society – in intentional communities, ecovillages, permaculture settlements, urban communes, rural religious and secular retreats, co-housing projects and activist squats.

My intention in this book is to bring together insights gained from meeting people in alternative settlements with those gained from utopian histories and theories, so that each can be understood in ways grounded in the other. I offer no definition of an ideal utopian society, nor of its appropriate built architecture, because the lesson of practical utopian or alternative settlements is that they are not ideal but real places, where much of the daily practice of a new society consists of finding ways in which people can collaborate on an equitable basis, respecting rather than obliterating difference. Utopia is thus incomplete, an always unfinished project. A few pervading questions nonetheless arise: the relation of the imminent (long or incremental) revolution of a society's culture and institutions to the immanent (immediate and all-pervasive) revolution of social transformation; how the envisioned new society comes into being; and the relation of change in social, economic and political structures to that of individual consciousness. I aim merely to read these as potentially creative tensions, so as to keep discussion open. Solutions are not my trade: historically, the more final they have been presented as, the more dangerous they have been.

Aims and scope

I wrote this book to gain a deeper understanding of utopian histories, stories, theories and practices. I do not live in an intentional community: my community lies more in the dispersed networks offered by my academic work rather than in any geographical roots. My thinking is done in the course of reading and writing, and I have minimal aptitude for field work. Nonetheless, I was drawn to the subject-matter from commitment to the idea that another world is possible, and realization that parallel to the available literatures in several academic fields there is the evidence of alternative modes of dwelling. Whatever the contradictions of the idea of Utopia there are, then, sites in which the difficulties are being addressed. This appeared a viable research programme, involving understanding of the role of literary Utopias as social critiques, the utopianism of modern thought and culture, the contribution of theory to the realization of a better world, and of how contemporary utopian or alternative settlements work. Early in the process of organizing research for the book it became evident that while some alternative settlements occupy buildings that have been specifically designed or adapted for an alternative way of living, for example by enabling low-impact living, the social architectures of alternative settlements – the processes through which principles are translated into action, and a group maintains its cohesion – often underpin the means by which built architectures are designed, and are a basis for building, metaphorically, a self-organizing society.

My purpose is to introduce the material of literary, theoretical and practical Utopias to readers from a range of academic fields. I make connections between the literatures of different fields, and between these and the accounts of people living in what I take to be an alternative society. The claims I make for the book are modest, but I hope it will contribute to interest in a utopian content in cultural and social studies, and to a grounded approach in the emerging field of utopian studies. The material is at times unavoidably complex but I have done my best to write the book accessibly, so that readers in one field can access the literatures of other fields and practices.

My material is drawn from readings of utopian texts from the sixteenth to twenty-first centuries, and from visits to alternative settlements in India, Europe and North America, and to historical sites in Britain and the United States, as well as to Egypt. Limited time and resources required me to be highly selective in the sites I visited, as did practical issues such as whether I could contact someone to arrange the visit or, as a non-driver, whether I could reach the site. I sought a geographical balance, and contrasts in the location, scale and types of settlements. It was necessary, too, to be selective in reading (and re-reading) texts from a wide range of literatures, and I have followed a path of personal choice inasmuch as I have followed up such connections and allusions as arose during the course of the research. The book is thus far from comprehensive in its coverage, though I have avoided a narrow definition of Utopia by including specifically utopian texts beside those dealing in a broader way with radical social change (but which may not point to a definitive ideal world). I hope I have not reproduced the habitual modern dualism of Utopia as ideal and social life as real. My original title for the book was *Building Hope*, the point of which is that the new arises within the old, and at a certain moment becomes radically different.

Structure

The book is structured in four parts. Chapters One to Three reconsider literary and theoretical Utopias from the early modern period to the nineteenth century. Chapters Four to Seven investigate recent and contemporary urban utopianism, focusing on social and environmental issues, activism and ecovillage living. Chapters Eight and Nine look to wider horizons in practices in the non-affluent world. Part Four consists of nine short, contrasting case studies aiming to give glimpses of alternative settlements – one historical and eight contemporary. This is followed by a conclusion in which I summarize the book's main arguments and revisit the problem of why Utopia does not dawn even though it is now technologically possible.

Chapter One reconsiders the literary Utopias of early modern literature; Thomas More's *Utopia* is read as a critical rather than prescriptive text. Chapter Two reviews aspects of René Descartes' search for certainty in the context of doubt and social upheaval, reading his work as informing a utopian strand in modern thought. Chapter Three examines Charles Fourier's plan for a total reorganization of French society, comparing his ideas to those of Robert Owen and the Chartists. Moving to the twentieth century and the practical dimension of utopianism, Chapter Four inquires into the planning of new cities, contrasting the model of the garden city to Le Corbusier's idea of a total erasure of the past. Chapter Five then investigates social Utopias from the 1960s to the 1990s, in counter-cultural revolt and a move to intentional communities. Chapter Six outlines frameworks of environmentalism, and the kinds of knowledge produced in eco-social formations, as prelude to accounts of anti-roads activism and ecovillage living in the 1990s in Chapter Seven. Chapter Eight deals with Hassan Fathy's mud-brick architecture from the 1940s to the 1960s as a possible alternative Modernism, and notes cases of alternative approaches to housing in the non-affluent world. Chapter Nine describes the work of the Social Work Research Centre, or Barefoot College, in Rajasthan, India, a rural campus for villagers' empowerment.

The nine case studies are based on visits to specific sites. The first is a historical site, the village of Economy, Pennsylvania, founded in the early nineteenth century by a Lutheran sect. The next two are intended as experimental cities: Arcosanti, a hilltop settlement in Arizona; and Auroville, an international city emerging from an ashram in Tamil Nadu, India. These are followed by three cases of autonomous settlement: the free city of Christiania, Copenhagen, founded by squatters' occupation of a disused barracks; Ufa-Fabrik, a cultural and ecological commune in Berlin, occupying a redundant film studio; and the district of Uzupio in Vilnius, Lithuania, a cultural quarter proclaiming itself as an autonomous zone. Finally there are three contemporary alternative settlements: Cambridge Cohousing, Massachusetts; the Ecovillage at Ithaca in New York State; and Zentrum für Experimentelle GesellschaftsGestaltung (Centre for Experimental Culture Design) at Belzig, Germany, an ecovillage developing the means for self-organization, conflict resolution and work on love and sexuality.

In the Conclusion, which is far from conclusive, the question of why Utopia remains a distant (or distanced) prospect is re-addressed, as is the perennial issue of whether social or individual liberation is prerequisite to a new society. This draws on the ideas of Henri Lefebvre, Herbert Marcuse and Hannah Arendt, opening questions for further research rather than resolving the always incomplete work of utopian thought.

Part One
Histories and theories

1 Imagining places
Literary Utopias and the far-away

Introduction

This chapter considers the literary Utopias of early modern Europe, focusing mainly on Thomas More's *Utopia* (1516–17) and dealing briefly with Tommasso Campanella's *City of the Sun* (1623) and Francis Bacon's *New Atlantis* (1629). All three authors set their narratives in distant geographical sites, and More and Campanella use the device of presenting their narratives as tales heard from fictional others, while Bacon tells the story of a sea voyage from Peru to China in the first person plural, echoing the convention of the genre of travellers' tales. Distancing allows a criticality which might in a more overt form have endangered the writer's liberty. It also allows a distinct, imaginative realm to emerge in which to construct a counter-image of the reality of the writer's own society. Yet the distancing of a utopian world from reality separates the ideal from the real, and this in time becomes a convention in which the utopian world is viewed as unattainable except in literary texts or visionary dreams. In reality, there was a parallel current of millenarianism and popular insurrection seeking to proclaim a new world in the here and now; so utopian ideals ironically become the reassuring foil to social rupture, in that the idea of a better world is set in a not-to-be-attained future that in turn tends, as a lost Eden or Arcadia, to be based upon the ideal of a remote and unknown place. This shift occurs over the modern period as a whole but appears inherent in the critical distancing that is a key characteristic of More's text; it is a flaw that haunts the whole history of the utopian imagination and begins in what can reasonably be claimed as its foundational text, *Utopia*.

I approach these issues from a reading of More's text, after outlining two contexts for its writing, one of upheaval in early modern England, the other of the evolution of the modern subject in early modern drama. I discuss More's text, then Campanella's and very briefly Bacon's. Finally, I note a second genre of utopian writing in the eighteenth century in which an ideal state of society is projected not onto far-away unreal places but onto the real far-away sites of European colonization, such as the South Seas. In the work of Jean-Jacques Rousseau, earlier in John Neville's *The Island of Pines* (1668) and Daniel Defoe's *Robinson Crusoe* (1719), or later in Herman Melville's *Typee* (1846), the far-away is represented as the site of an uncorrupted humanity, an Eden that never was and becomes a literary or aesthetic entity.

Insurrections

For More, the gap between narrative and reality reflected a need to negotiate a path between contesting factions of power in the English state. The early modern period in England saw repeated uprisings, usually beginning among a displaced rural poor in a period of land enclosures that dispossessed them of the common lands for grazing and foraging on which their marginal livelihoods depended. This current fed, and was in turn fed by, a growth of radical religious sentiment revolving around the personal, unmediated interpretation of scripture. The Lollards, for example, preached direct, personal reception of scripture and reliance on inward religious feeling rather than ritual observance. Such movements challenged the authority of church and state in an assertion of the rights of common people, for the most part when they had almost nothing left to lose and took to the road, and were ruthlessly suppressed. In 1417, a planned Lollard uprising was foiled when their leader, Sir Thomas Oldcastle, was arrested, brought to London and executed. This was 99 years before the publication of *Utopia*, but handwritten Lollard bibles were still in circulation in the 1500s, and the currents of this folk religion, so to speak, which was also an attempt to regain access to common land, continued to ebb and flow through the sixteenth century, eventually erupting in another form in the radical sects of the English Revolution in the 1640s. The point here is that this current forms part of the context for More's writing. It connects medieval millenarianism, in which the new age was often associated with radical social transformation – as in the ending of toil and abolition of office and property – and the Peasants' Revolt of the 1380s (Fremion, 2002: 17–22), with the Diggers and the Levellers in the 1640s (Hill, 1975; Petegorsky, 1999), and perhaps even with nineteenth-century utopian socialism and anti-capitalist direct action in the 1990s. In the immediate context of More's *Utopia*, what is often taken to be a medieval trend to uprising (Kohn, 1970) remained a real force, while popular anti-clericalism was harnessed by Henry VIII as useful to his own aim of autonomy from Rome (identified as the seat of corruption in the church). In fact, Henry's appropriation of the lands and wealth of the monasteries exacerbated the conditions of the rural poor for whom the monasteries had provided a safety net, thus increasing the likelihood of revolt (Coates, 2001: 12).

For groups like the Lollards, or the Anabaptists of Münster in the 1530s (Midelfort, 1999; Kohn, 1970: 261–71), the certainty that the new age, guaranteed in revelation, was about to dawn compensated for any material risk in insurrection. If the old world was soon to end, a belief likely to be strengthened by worsening material conditions, and if insurrection was the means to usher in the new (already present in consciousness, and perceived as soon to come), those who regarded themselves as the Elect were safe in their impending salvation and were required, as instruments of revelation, to revolt against those, usually in the elite, whose very presence stayed the new age. In July 1524, for instance, Thomas Müntzer preached, 'Drive Christ's enemies out from among the Elect, for you are the instruments . . . The sword is necessary to exterminate them . . . At the harvest time one must pluck the weeds out of God's vineyard . . .' (quoted in Kohn, 1970:

239). This was eight years after the publication of More's *Utopia*, and seven after Martin Luther nailed his theses to a church door in Wittenberg.

More, as a state servant and loyal Catholic, had no dealings with insurrectionists. Nor could he make overt criticisms of church or state – such freedom of expression was not a known idea. Yet he sailed close to the wind, and his text was published in Louvain because, as Henry Morley says, 'its satire was too direct to be misunderstood, even when it mocked English policy with ironical praise for doing exactly what it failed to do' (Morley, 1887: 7). In the end More was executed for putting faith before secular authority, but this does not mean *Utopia* is a radical text of a kind similar to the political pamphlets that circulated in the 1640s. It was, however, a critical text, and in one way *Utopia* might be read as a critical account of social disorder in England in the 1510s parallel to the undercurrent of unrest; while in another, to which I incline, it might be read as affirming that state of affairs by distancing its critical viewpoint to a traveller's tale, an incidental, anecdotal yarn. In both cases, which are compatible, the society described at a geographical distance in *Utopia* is emphatically not a prescription for what England should become but, as I explain below, an oblique mirroring of what it was – through which representation to inform an incipient discourse of the English state.

Having said that, More's text did inform later radicalism in which the aim was to introduce a degree of social equity that was unforeseen in More's time. Christopher Hill describes how *Tyranipocrit Discovered*, an anonymous text printed in the Netherlands in 1649, echoes More's text in denouncing the rich who hang the poor when it is the acts of deprivation perpetrated by the rich themselves that force the poor into theft in the first place (Hill, 1975: 116). The image of thieves hanging from the gibbet is a motif in More's story, presented with an almost identical argument in the opening section of the text: a seafarer called Hythloday whom More meets in Antwerp criticizes the disorder of a society in which people are forced into theft and then hanged by the dozen while others, equally poor and deprived of livelihood, in turn become thieves while, all the time, the true thieves, as it were, are the rich who steal the poor's livelihood. More, of course, must distance himself from such scenarios, setting *Utopia* in a far-away place and putting the narration in the words of a traveller whom he meets inconsequentially, the conversation that ensues being an informal exchange incidental to, almost a distraction from, More's service as a diplomat and state servant (in which role he went to Antwerp). To play with voices, though, indicates a reflexive idea of a speaking subject.

Context – the modern subject

Utopia begins in More's text as an image of somewhere else, used as a critical device through which to write about the author's society by other means in a time of censorship. Set in the author's own time but not the author's place, displacement of the narrative to the words of another person further removes the text from present reality. Utopia, that is, exists only in verbal description. More clearly uses this as a safety device, combined with wit and, as I outline below, a coded aspect designed for an elite readership among his own milieu. But the narrator also relays other voices, just as the structure of the text contains stories within the story. This reflects, and is part of, the development between the sixteenth and the nineteenth centuries of the idea of a human subject (self), the author whose thought is articulated as literature. The idea of a self which speaks itself is tentative in *Utopia*. Its modernity lies primarily in the way that More relays an original story rather than interprets previously received knowledge (as in the medieval tradition of scholasticism). The story is not his own but someone else's, which he retells from interest or amusement. It might even be a joke or a joker's tale, if need be. The self that relates the tale of a tale has now become complex by reflecting on its own location, able to move at will from one character to another.

The modern self made in its own perception of the subject is evident, as Catherine Belsey shows (1985), in early modern drama: first in the sixteenth century, performed on what was still a stage in the round; then, in the seventeenth century, extended in a move to performance on a proscenium stage on which actors remain driven by forces such as desire, power and the perceptions of others but seem to make choices that are their own. From this development Belsey identifies the subject of liberal humanism who possesses agency, emerging in the seventeenth century, the century of the English Revolution and as 'an effect of the revolution . . . when the bourgeoisie is installed as the ruling class' (Belsey, 1985: 33–4). Belsey contrasts the medieval subject, having no unifying essence – body and soul, mortal container and immortal content awaiting the Day of Judgement – with the modern subject in Shakespeare's *Hamlet* (1601), for whom the world is a stage on which to perform the self called Hamlet. Yet the binary model of the medieval subject persists – the actor 'is another paradox: like a god and at the same time a quintessence of dust' (Belsey, 1985: 27). Hamlet speaks a choice between being and not being, in a statement of agency which includes the agency to silence the speaking self. Raymond Williams notes that the soliloquy – the speech to oneself characteristic of early modern drama – rehearses the potential for a concept of the individual (Williams, 1981: 142, cited in Belsey, 1985: 42).

Hamlet is not the only such character in Elizabethan or Jacobean drama. Belsey writes of Christopher Marlowe's *Doctor Faustus* (1592?) that the soliloquies denote a genealogy, 'rendering precarious precisely the unified subjectivity which it is its project to represent . . . [when] the repressed discontinuities of the allegorical tradition return to haunt the single voice which speaks' (Belsey, 1985: 44). She adds, citing Thomas Kyd's *The Spanish Tragedy* (1590) and Shakespeare's *Macbeth* (1606), that a gap opens up now between the representation of a subject and the interiority presented in soliloquies: 'the "I" cannot be fully present in what it says of itself' (Belsey, 1985: 49), to reveal an identity behind the words. If the liberal-humanist subject is ambivalent in its location, 'the unified and unique subject of liberal humanism is forever tragically locked within its own silence' (Belsey, 1985: 52). All this post-dates More's text but I make this apparent detour because it may illuminate the ambivalence of the speaking subject in *Utopia*, a subject establishing a critical distance that is itself one of the stories in the story, and a story of modernity.

Stories in stories

If More's text is contextualized by an emerging concept of the subject as possessing agency in a period when power was in the hands of princes and prelates, then, and by the capacity of that emergent subject-self to reflect upon its own being, it is also an effort to deploy voices as elements of critical distance – a play on literary devices. The comparison of ideas in the narration is expressed through the voices of characters in the story, one of whom tells his own story but is not a real person while another, narrating this, is the real author who has no position of his own, only the story he passes on. Ernst Bloch glosses, 'the dream of the best state is here once again presented as a kind of sailor's yarn' (Bloch, 1986: 516), reading *Utopia* as a proto-bourgeois text. But Peter Carey notes (1999: 38–9) that More allows doubt as to his seriousness by distancing himself from the text. Carey points out that the name Hythloday means dispenser of nonsense or windbag. Utopia means no-place, and if the society described is blissful this may be either, Carey argues, a way to consign it to the inconsequential or, as double-bluff, a simulation of a joke, 'in a society where spreading new ideas could be dangerous' (Carey, 1999: 39). Satire was in fact not uncommon in the period, but More's text is transgressive in two ways: the story told by Hythloday which More retells; and the implicit story of More's own negotiation of a space in which to speak critically. The implicit signs of transgression include the fact that Hythloday is a seafarer who thus crosses the boundaries of states, his realm the free and formless ocean; and that More meets him in the port of Antwerp, a node of printing and trade where radical texts were known to circulate. More crossed state boundaries, too, as a diplomat and member of a new, literate, administrative class, one of the functions of which was to oversee the organization of trade and the collection of tariffs. More adopts a disguise, then, as the narrator of a story, and frames it such realistic detail that it becomes plausible. His independence from the story's author meanwhile absolves him of the story's burden of criticality.

The realism rests on accounts already published of voyages seeking to circumnavigate the globe. Hythloday's voyage is a sub-category, then, of Amerigo Vespucci's account of his voyages in 1499–1503 (published in 1507). Vespucci left 24 men at Cape Frio in 1504 – the starting point for Hythloday's fictional voyage. The traveller's tale became a literary genre in its own right during the next three centuries, lending itself, as Louis James notes, 'to utopian ideals' (James, 1993: 34). James notes that over 500 island stories appeared between 1788 and 1910 (James, 1993: 39, citing Carpenter, 1984).

Antwerp

The first section of *Utopia* describes More's meeting with Hythloday while carrying out a diplomatic mission for Henry VIII in 1515 (which More did actually undertake). This sets the narrative in an environment of exchange among cosmopolitan equals in literacy: More is English; Hythloday is Portuguese, was with the Italian Vespuccci in New Spain, knows some Latin and reads Greek. He is introduced to More by Peter Gilles, (actual) town clerk of Antwerp, 'both a very worthy and a very knowing person, so he is civil to all men' (More, 1997: 1). Also introduced in the narrative are John Morton (past Chancellor of England), a group of English lawyers and a jester. Taking the trope of the hanging of 21 thieves to a gibbet in England, Hythloday, in a conversation with John Morton (hence not directly with More, who merely happens to be present), says the punishment is too severe and in any case ineffectual. Poverty is the cause of theft and ensures that however many thieves are hanged there will be more because 'what else can they do? For . . . they have worn out both their health and their clothes, and are tattered' (More, 1997: 7). Hythloday continues that land enclosures by the rich and powerful cause the poverty which leads to crime: 'The increase of pasture . . . by which your sheep . . . may be said now to devour men, and unpeople, not only villages, but towns' (More, 1997: 9). Further complaints concern luxury, drink and gambling. In answering a question as to what punishment he favours in place of death, Hythloday mentions that among the Polylerits – a society in Persia – restitution of property is made to its owner, 'and not as it is in other places, to the prince, for they reckon that the prince has no more right to the stolen goods than the thief' (More, 1997: 11). After several excursions into the ordering of a society, More, Gilles and Hythloday go to dinner, after which Hythloday describes Utopia, which constitutes Book II – an after-dinner entertainment.

More anticipated that his readers would understand a coded nomenclature, and would have a keen interest in issues of property, wealth and the state, which implies his pseudo-realism is a bluff. A utopian alphabet is appended, with a logical system of inflexions, to add to the effect. Paul Turner writes (introducing the Penguin edition), 'One function of these realistic devices is to amuse intelligent readers by hoaxing the unintelligent . . . This pretence of second-hand reporting may be regarded as another protective technique' (More, 1965: 9–10). More's interest was not in rebellion but in a sense of a spiritual community, and Elizabeth Grosz writes that the calm sea that surrounds Utopia is an emblem of political harmony while 'the Utopians live in the best form of commonwealth, though one with its own terrible costs, the intense constraints on personal freedom' (Grosz, 2002: 266). Her comment could indicate the constraints, but also the peace, of a monastic community. The Polylerits wear quasi-monastic habits of one colour, and have cropped hair (More, 1997: 13) and for Bloch this points to More's interest in early Christian asceticism. Aligned to the English state's perception of corruption in the church this suggests that the target of More's coded narrative might perhaps be Rome. Bloch writes, 'it is not England that is supposed to have been criticized . . . but the Pontifical state.' Leaving aside Bloch's claim that the text is not More's original but an adaptation by Erasmus of Rotterdam, Bloch's assertion that Utopia is Mount Athos, a Greek Orthodox monastic community on an island with a rocky coast, is not uninteresting: 'Instead of England and the primitive commune [of American settlers], More is solely supposed [in the account of Heinrich Brockhaus] to have contrasted the two centres of Christianity: Rome and Athos; Utopia is . . . Athos reconstructed' (Bloch, 1986: 517). Bloch reads this as an appeal for church reform, yet More was a devout Catholic prepared to put faith before state, which seems to deny Bloch's interpretation. Perhaps more central to the text, then, is Hythloday's vehement criticism of injustices in England; yet here, too, there is a possible ambivalence inasmuch as images of material corruption and dis-ease prompt attention to the immaterial harmonies of the spheres and the order and perpetuity of the religious life. Further, very speculatively, an aspect of early modern aesthetics concerns images of the unbeautiful – of which hanged men are a primary case – that can be rendered beautiful in the quality of representation, while acting as foils to images of a more direct beauty. For example, two hanged figures are present in the background of Pisanello's fresco *St George and the Princess* (1433–8, Verona, St Anastasia), in which the two noble figures stand either side of a horse's bottom. Of course, the sight of hanged thieves was not uncommon in early modern Europe, and in London the heads and body parts of those executed for treason were publicly displayed.

The island of Utopia

Utopia is an island surrounded by a serene sea and guarded from visitors by a rocky foreshore and a port to which ships need piloting: 'the coast is so fortified, both by nature and by art, that a small number of men can hinder the descent of a great army' (More, 1997: 28). Hythloday is precise in his geography. The island is 200 miles wide. The river in Amaurot, the main city, has a tidal reach of 30 miles and its streets are 20 feet wide. The precision adds to a realism designed for one group of readers, while the name Amaurot means castle in the air and the river's name, Anydrus, means no-water, which an elite would have understood: it is as if More sets his readers an intelligence test. He cites Utopia's administration: 'There are fifty-four cities in the island, all large and well built: the manners, customs, and laws of which are the same . . . The nearest lie at least twenty-four miles distance from one another' (More, 1997: 29); each city is no more than a day's journey on foot from the next while each sends three senators once a year to Amaurot, 'to consult about their common concern' (ibid.). The jurisdiction of each city extends to the mid-distance between cities so that 'no town desires to enlarge its bounds, for the people consider themselves rather as tenants than landlords' (ibid.). More depicts the Utopians as inclining to cerebral rather than sensual joys:

> But of all pleasures, they esteem those to be most valuable that lie in the mind; the chief of which arises out of the true virtue, and the witness of a good conscience. They account health the chief pleasure that belongs to the body; for they think that the pleasure of eating and drinking, and all the other delights of the senses, are only so far desirable as they give or maintain health. But they are not pleasant in themselves, otherwise than as they resist those impressions that our natural infirmities are still making upon us: for as a wise man desires rather to avoid diseases than to take physic; and to be freed from pain, rather than to find ease by remedies; so it is more desirable not to need this sort of pleasure, than to be obliged to indulge it.
>
> (More, 1997: 53)

Again, the monastic life comes to mind. Grosz comments that More follows Plato's *Republic* in devoting considerable space to the organization of sexual relations and relative status of men, women and children (Grosz, 2002: 273). In Utopia, both genders work alongside each other, and their productivity means that the working day is only six hours long. But the rules for marriage, divorce and sex 'are strict and govern a narrow, life-long personal and non-deceptive monogamy' (ibid.). Prospective marriage partners – minimum age 18 for women and 22 for men – are shown naked to each other prior to marriage in an arrangement likened to the purchase of a horse (More, 1997: 58–9). Divorce is allowed, but infidelity is punished by slavery (into which injured partners may follow if they do not wish for separation). A further religious allusion is made almost in passing in the lack of visible distinction of the Prince, 'either of garments, or of a crown; but is only distinguished by a sheaf of corn carried before him' (ibid.: 61), and of the High Priest – simply preceded by a person carrying a candle. The utopian calendar has a pagan or archaic aspect: 'The first and last day of the month, and of the year, is a festival. They measure their months by the course of the moon, and their years by the course of the sun' (More, 1997: 78). There are temples in the dark interiors of which a Divine Essence is worshipped by followers of many religions, and More continues with an iconoclastic reference: 'There are no images . . . so that every one may represent Him to his thoughts . . . nor do they call this one god by any other name but that of Mithras' (More, 1997: 79). All except the priests wear white at festivals; wives and children confess their sins, 'thus all little discontents in families are removed, that they may offer up their devotions with a pure and serene mind' (ibid.); and the sexes are separated, men on the right and women on the left, as indeed they were when lay people went to monasteries to hear mass.

The basic social unit in Utopia is the family. Each city is allowed to accommodate up to 6,000 families, so that none becomes too large or small. The size of families is also moderated to between 10 and 16 adults, children from over-abundant households being placed in those with few. Female children are married out but males remain in the family house. Within the family, the eldest male is the head of the household. So, 'wives serve their husbands, and children their parents, and always the younger serves the elder' (More, 1997: 38). Today this reads as a model of patriarchy inadmissible in a gender-equal society. But it may not have seemed unusual in Tudor England, and the detail of More's structuring of Utopia acts not only as description but also as literary form, a model of orderliness to provoke contemplation of present disorders. Several sixteenth-century English texts denote at least a fear of social disintegration, among them Elyot's *The Governor*, Hooker's *Laws of Ecclesiastical Polity* and a church homily titled *Of Obedience*. Elyot writes, 'Take away order from all things, what should then remain? Certes nothing finally, except some man would imagine festoons chaos' (cited in Tillyard, 1963: 19). For Hooker, church law follows natural law, or rather God's laws produce natural order articulated as a chain of being from the deity and the angels to humans, beasts, plants and rocks. This world, or cosmos, is ordered by degree, the non-observance of which is chaotic (Tillyard, 1963: 22). These texts post-date More, but the tendency to order and degree is common to them and *Utopia*. More ends *Utopia* by saying that while there are many interesting and useful aspects to Hythloday's story, 'I cannot perfectly agree to everything he has related' (More, 1997: 85). In the penultimate paragraph, Hythloday ends his story by affirming utopian society and its laws which 'will never be able to put their state into any commotion or disorder' (More, 1997: 84), a literary foil to the state of early modern England.

Re-reading More

As indicated in the contextual material above, More's England witnessed contests for social justice as the conditions of the rural poor worsened. In 1536, the Pilgrimage of Grace began in the north of England, seeking to end land enclosures and restore the old Catholic holidays. In 1549 a rebellion occurred in Norfolk, led by Robert Kett, and was able to destroy enclosure fences around the city before the arrival of 12,000 soldiers to crush the revolt – a massive military presence in that period. The pattern of reclaiming the right to land continued when the Diggers dug up St George's Hill in Surrey on 1 April 1649 (Coates, 2001: 21–5; Petegorsky, 1999: 153–76). Elsewhere in Europe, the Anabaptist Free Commonwealth was established in 1534 in Münster, where John of Leyden preached common ownership of property and sexual freedom. Earlier, in 1528, a group of religious refugees fleeing to Austerlitz proclaimed the common ownership of property (Hostetler and Huntington, 1967: 2). It appears that Hythloday's reference to sheep devouring villages is far from fanciful, and his declaration of a perpetual freedom from disorder acts as foil to the reality of common life.

There are several difficulties: Thomas Markus writes that nothing obscures 'utopia's double effect: oppressive, alienating elements coexist with liberating, humanizing ones' (Markus, 2002: 15). But the attractive permanence of perpetual freedom set out as regulation on quasi-monastic lines is also a rigidity which in today's reading of the text lends utopianism the negative aspect of an inability to accept change, a suffocating lack of adaptability. There is a suppression of difference, too, Grosz arguing that 'the question of sexual difference has not been raised' (Grosz, 2002: 274) and that the partial egalitarianism of *Utopia* is not a structural rethinking of rights. Grosz continues:

> At best equal participation is formulated. But the idea of sexual difference entails the existence of *at least two* points of view, sets of interests, perspectives, two types of ideal, two modes of knowledge, has yet to be considered. It is, in a sense, beyond the utopian, insofar as the utopian has always been the present's projection of singular and universal ideals, the projection of the present's failure to see its own modes of neutralization. Sexual difference, like the utopic, is a category of the *future anterior*, Irigaray's preferred tense for writing, the only tense that openly addresses the question of the future without, like the utopian vision, pre-empting it.
>
> (Grosz, 2002: 275)

There is a sense, too, that utopian desire attaches to the unattainable as its object. For Michael McKeon, imaginary voyages combine 'a recognizable progressive critique of contemporary social stratification with a nuance of conservative doubt concerning the practicality of attaining utopias' (McKeon, 1987: 7, cited in James, 1993: 34), and Amy Bingaman, Lise Sanders and Rebecca Zorach argue that utopian principles can be actualized in corporeal life only as contradiction – 'the body signifying unpredictability, concreteness and change, and utopia characterized by predictability, abstraction and permanence' (Bingaman, Sanders and Zorach, 2002: 2).

Nonetheless, for Bloch, 'The "Utopia" is and remains, with all its dross, the first modern portrait of democratic-communist wishful dreams' (Bloch, 1986: 519). This leads me to ask how More's text represents a realm of the not-yet but attainable. If it is taken as prescriptive, much of it would be rejected by most progressive commentators now. Carey, for instance, summarizes the society described in *Utopia* as irking today's readers while possibly being attractive to More, who wore a hair shirt under his state robes. He writes, 'Clothing is uniform, and made of undyed homespun wool. You need a permit to travel, and must go in a group. If you travel without a permit, you are arrested as a runaway . . . Marital arrangements are strict and unsentimental' (Carey, 1999: 39). It sounds like a travel guide for an authoritarian state. But *Utopia* is a made-up place, either serious or a joke, either prescriptive or provocative in the degree to which what it describes differs from the conditions in which its readers know they dwell. It thus offers no gap between description and actuality, in contrast to the charge that the utopianism of state socialist regimes (presented in Carey's terms as uniform and frugal) is precisely the difference between revolutionary dream and reality. I want, then, to insist on the literary status of More's text: a story within a story, a text within a text, the coded imagery of which is a vehicle for a criticality that could not reasonably have been articulated in any other way. No other way would, in any case, have appealed to More's readers, able to understand his coding of a pseudo-reality. Judith Shklar writes: 'utopia was a model, an ideal pattern that invited contemplation and judgment but did not entail any other activity' (Shklar, 1973: 105). Or perhaps *Utopia* was a jest in the sense that the jester's licensed jibes provoke judgements otherwise inadmissible: 'There was a jester standing by, that counterfeited the fool so naturally, that he seemed to be really one' (More, 1997: 15). More does not play the fool. The relation of the fool to the company he addresses remains ambiguous, like More's relation to his text.

Campanella's text

The island of Utopia is occupied by evenly spaced cities. Though it has a chief city, it is de-centred like a monastic realm of more or less equally important settlements; Campanella's image of a City of the Sun, by contrast, indicates a highly centralized, as if urbanized, world in which the city is like a sun in the landscape, all roads leading to it. Campanella was a Dominican, imprisoned for 27 years following implication in a conspiracy against Spanish rule of Naples, during which he wrote *The City of the Sun* (*Civitas Solis*). The work is subtitled 'A Poetical Dialogue between a Grandmaster of the Knights Hospitallers and a Genoese Sea-captain, his guest'. Like *Utopia*, it uses the literary device of a seafarer's tale.

The city Campanella describes is radial, divided into seven rings named after the seven known planets and intersected by four gated streets aligned to the points of the compass. Initially this is represented as a defence, each ring a self-contained defence-work around a hill on which the centre of the city (its seventh, innermost ring) is a domed, circular temple at the summit. The city is a theocracy ruled by a high priest named Hoh (which Campanella translates as Metaphysic), assisted by three princes whose names – Pon, Sin and Mor – indicate power, wisdom and love. But love is not a matter of personal choice, rather (as in More's text) a facet of life regulated for communal well-being: love born of desire is unknown, and 'the rules regarding procreation are observed religiously, for the public good' (Campanella, 1981: 32). As in *Utopia*, the young respect the old but in the City of the Sun there are no slaves, only citizens dressed uniformly in white who work willingly and cooperatively. Wisdom oversees the liberal arts, including the depiction of all knowledge for public education. Each circuit of walls carries on both its inner and outer faces a sub-set of an already-complete knowledge: first, mathematical figures and the Earth; second, minerals and geographies; third, flora and fish; fourth, birds and crawling beasts; fifth, animals on both sides; sixth, mechanical arts and their instruments, and figures such as Osiris and Mercury credited as inventors. I compress Campanella's description to emphasize the model of completeness, a finished society and its finished knowledge. The solar city again differs, then, from a transposition onto a geometric space of a monastic order, in that knowledge is made public, not enclosed in a scholarly community which guards its Latin texts and hence safeguards its interpretation; each generation of the city's population is taken round the circuit of the murals in their education. It is also like a planetary model, the planets being depicted in the decoration at the centre of the temple, the city and knowledge as complete as the solar system (assumed to be immutable), while references to the zodiac occur several times in the text.

For Bloch, Campanella draws his imagery from the Inca state in New Spain. Bloch also notes Campanella's dedication of a republication of his book to the French Cardinal Richelieu in 1637: 'devised by me to be established by you' (Bloch, 1986: 524). In the conquest and white settlement of the Americas, new cities were preconceived in the form of an emblematic plan drawn by a military commander or a Viceroy (Low, 2000: 96–9) but, contrary to Bloch's idea, the colonial city of the Americas was in most cases a grid city with perpendicular main streets and a large central open square (*plaza major*), and derived from types such as the Mexican city of Tenochtitlan. The regulation of colonial space was codified in 1573 in Phillip II's Laws of the Indies, and the writing of utopian texts can be seen as a parallel activity – the one codifying the material space of conquest; the other an intellectual space of criticality. The specification of spaces and proportions is a key aspect, an idealized element possible in the exercise of power or the authoring of a text. I say more of the this in Chapter Two, seeing it as a quintessentially modern act.

From New Atlantis to the noble savage

If Campanella's ideal city does not correspond to the cities built in the Americas, and has the feeling more of a cosmological diagram, it nonetheless denotes a transition from the de-centred world of *Utopia* to a centrally planned world, in which knowledge is the guarantor of unbroken continuity. But for Bacon, in *New Atlantis*, knowledge is also the source of power over the natural world and other human subjects.

The known world, as it geographically extended in the seventeenth century across a globe known to be circumnavigable, and as it intellectually extended to the codification and ordering of wild nature and human nature, was disenchanted. I mean by that, taking my idea from critical theory in the 1940s, that it was no longer subjected to the rule of a mysterious Fate but was, even if a divine creation, reproduced by human intervention in nature. Rocks and springs no longer harboured spirits but were substances and processes to be weighed and measured, then controlled and perhaps improved. Similarly, cities were mapped, at first in order to be taxed, leading to a rapid expansion of map-making in the seventeenth century (Pickles, 2004: 92–106); but the device of the map then allowed the planning of a city that had not hitherto existed, as if on a blank space. Anything could be done. As Theodor Adorno and Max Horkheimer say in *Dialectic of Enlightenment*, 'the human mind, which overcomes superstition, is to hold sway over a disenchanted nature. Knowledge, which is power, knows no obstacles' (Adorno and Horkheimer, 1997: 4). At the same time, the dissolution of myth in science leads, by an oblique route, to the reinstatement of myth in new forms of scientific and institutional convention. Rationality, then, contains plural currents, one of which is instrumentality – the model that a specific action achieves a specific and repeatable end – or the institutionalization of that power over others and the natural world first perceived by Bacon.

It is a patriarchal undertaking: the model of a masculine power-over is reproduced in colonial power when the colonized are categorized as primitives, closer to a natural realm of urges and necessities (just as women were seen as more subject to irrational desires than men), hence less humanized. In the nineteenth century this is translated as the white man's burden as a cover for resource extraction. But in the eighteenth century there is a philosophical resistance to such acts of intellectual, or textual (but not so much real), colonization of the inhabitants of distant places seen as far-way Edens. For Jean-Jacques Rousseau, the inhabitants of colonized islands in the South Pacific represent a humanity uncorrupted by luxury. If rationality expresses natural law and is a foundational quality to be regained after diversions and abuses collectively described as European civilization, then the type of the noble savage represents a personification of the pre-corrupted state. Prior to corruption by commerce and power, humans are for Rousseau essentially good. He elaborates this theory in *Emile* (1762), describing the education of a child. Rousseau foresees Emile as beginning to relate to the wider world at the age of 12, learning manual skills as well so that he can collaborate with others in making things. Rousseau regards the arts as harbingers of luxury and decay, a vehicle of decadence unless used for educational purposes, but prescribes for Emile a first literary work – Defoe's *Robinson Crusoe* – that will 'serve to test our progress towards a right judgement' (cited in James, 1993: 39). As James points out, Defoe's book had been rather dismissed when it was first published, but became 'a major European text' (James, 1993: 39) after Rousseau took it up as an educational work. The novel, like many utopian texts, is set on a remote island on which Crusoe is washed up after a shipwreck. It has a political dimension in that Crusoe returns to England just before the Glorious Revolution of 1688, his island stay possibly analogous to the restored Stuart monarchy.

For Rousseau, however, the shipwreck denotes the demise of European society and the island a return to an original innocence. James continues:

> By taking the island as a 'real' place, Rousseau paradoxically makes it utopian. For it envisions an environment without complex social relationships, free from the complications of sex and of the need to earn money. When Rousseau himself played Crusoe later in life, it was . . . to hide in a secluded island on Lake Brienne . . . Rousseau aligns Crusoe with the Romantic idyll, the natural state, an Eden that was to be envisaged in the *Paul et Virginie* (1788) of Rousseau's friend and translator, Bernardin de St Pierre, which is set on the island of Martinique.
>
> (James, 1993: 40)

Several objections are obvious: that Rousseau knew nothing of the real inhabitants of the South Pacific or Caribbean islands; that he abstracts and universalizes them; and that he ignores the real corruptions which were the donation of the Europeans to such peoples (in sexually transmitted and other diseases, alcohol, and later in guns). But the literature of a natural human state is a significant genre in the eighteenth century, as later in Herman Melville's *Typee* (1846). Melville describes the fictional escape of a group of sailors to a paradisiacal valley in Polynesia, citing Rousseau: 'the continual happiness, which so far as I was able to judge appeared to prevail in the valley, sprang principally from that all-pervading sensation which Rousseau has told us he at one time experienced' (Melville, 1972: 183). The inhabitants of Typee decorate themselves with garlands of flowers; men are busy fishing, carving or polishing ornaments; there is no strife, only perpetual repose as if every day is Sunday, the day of rest and enjoyment (except here there are Mondays on which to awake from the dream). Of course, Melville does not describe a material bliss any more than More, even though his overtly fictional narrative rests on a plausibility gained from real stories of colonial exploration. But by the 1840s, the idea of a world absolved of toil had taken on a politically revolutionary character.

Literary utopianism

To end this chapter I suggest that nineteenth-century texts such as *Typee* and many other island stories in European literature constitute a utopian literary tradition that begins in More's *Utopia*. Literary utopias are imaginative constructions of a world set far-away by which to draw critical attention to the world in which they are written and in which they are read. They reaffirm that another (or an other) world is possible outside the after-life of organized religion, but as an imagined other-world, thus not posited as a real-possible alternative society. The text may be a provocative device conducive to such thoughts of social reorganization, but also displaces those thoughts to an imaginative or aesthetic realm. That is the tension in which utopian writing is probably always held, as perhaps this book will demonstrate (itself a story of utopianism told through the utopian texts of others). This literary tradition stands apart from, and may be a foil to, that other historical strand of popular uprising, from medieval millenarianism to the radical groups of the English Revolution and anti-capitalist activism in the late 1990s. The difficulty is that the literature of Utopia depletes the force and urgency of a desire for a world free of oppression, sublimating it in order to postpone the utopian moment to a future as far away in time as the islands of utopian literature are far away in space.

Bloch differs from this line of argument, believing in an ever-present utopian content in art and literature, the role of culture being to articulate a hope that begins in dreams, is then shaped in daydreams and is given recognizable form in the arts. Recognition of the utopian content of art and literature then, as it were, brings Utopia within the grasp, to open a realisation of its object – the end of history – as freedom. This is not unattainable but objectively given. Does this square the circle by bringing the utopianism of high culture into the spheres of mass revolt? Bloch, alone in the circle of the Frankfurt School, reads popular culture as a positive manifestation of hope, not a means to dupe the mass public. He employs the idea of a Residual Sunday to denote the carefree, toil-free life of a utopian realm that he finds in both high art and folk tales. He sees it in Gauguin's paintings of Tahiti: 'Happiness and colour in Tahiti: this kind of thing now comes into the late Sunday picture, a long-lost piece of south land' (Bloch, 1986: 816), though the actualities of Gauguin's life in Tahiti contradict this – somatically in the effects of a sexually transmitted disease he acquired in Paris before his second voyage. Bloch borrows Gauguin, so to speak, to include him in a literary tradition in which the Land of Cockaigne in turn connects to millenarianism. It is an attractive idea, even if Bloch takes liberties of interpretation and is eclectic in his cultural references, though it is not necessarily a convincing one: the role of popular, as high, culture may be either to urge on revolt by producing a critical climate in which the dominant social structure becomes unacceptable in face of a mass critical awareness; or, as the aesthetic dimension is more often deemed to achieve (see Marcuse, 1968b), to defer revolt in favour of the luxuries of a literary island of delights. I leave the argument open at this point, but resume discussion in the book's Conclusion. I end here simply by juxtaposing an extract from Bloch's description of Gauguin's work in Tahiti and an extract from Francis Godwin's *The Man in the Moon, or a Discourse of a Voyage Thither* of c.1580. Bloch writes:

> a pair of children among flowers, a young rider and girls who await him, a Maori hose between palms, effortless fruit-picking. Everything lies in fiery brown repose . . . melancholy is dissolved or transformed into silence and

distant vacancy. 'Nature for Us', this most fundamental background of the Land of Cockaigne, is indicated by a paradisial region . . . The Sunday on Tahiti is bought with primitiveness, and this primitiveness is rewarded with carefreeness . . . But: terra australis *in Europe* . . . this workday as Sunday has not yet been painted anywhere; for it would be the classless society.

(Bloch, 1986: 817)

And Godwin writes:

Food groweth everywhere without labour, and that of all sorts to be desired. For raiment, housing, or anything else that you may imagine possible for a man to want or desire, it is provided by the command of superiors, though not without labour, yet so little that they do nothing but as it were playing, and with pleasure.

(in Carey, 1999: 48)

Both texts are fictions, though the continuity of a desire for work as play appears to support Bloch's idea that a latent utopian consciousness permeates all periods of human culture. That does not answer the question as to whether an image of a utopian realm is a means to its realization, or its perpetual deferral.

2 Drawing lines
Modernity and Utopia

Introduction

In the previous chapter I reconsidered early literary Utopias, noting the emergence of a context of social unrest, modern subjectivity, the use of devices to distance an author from a text, and the distancing of a utopian image to the far-away places of European colonization. I asked whether literary Utopias contribute to the realization of a radically reorganized society, or displace the desire for freedom to an aesthetic no-where, literally a u-topia. In this chapter I extend discussion from the literary to the philosophical, or more exactly to the metaphorical, focusing on an architectural metaphor used by René Descartes in his *Discourse on the Method of Conducting One's Reason Well* (*Discours de la méthode pour bien conduire sa raison*) of 1637. Descartes uses the image of an engineer tracing a line on a blank ground according to his imagination. I read this as a metaphor, though it could describe the work of engineers laying out new a town. But in the *metaphorical* act of drawing lines, the engineer defines an imagined form which has no material equivalent, inserting into the world a (Cartesian) space that did not previously exist but is made *in the act of drawing a line*. To extend the metaphor, this draws a line under the past as well, to separate modern knowledge from that of the past. I read this as indicating a utopian desire to make new that informs the planned cities of the Enlightenment and the twentieth-century project of engineering a new society by design. Implicit in the act of making new is the autonomy of the thinking subject, who has power to re-order the world in a shift from intention to plan that parallels the shift from literary Utopias to the projection of utopian images onto the real, thereby re-framed, sites of colonization.

The difficulty is that the observer is separated from the observed, ideas are privileged over practices, the ideal subordinates the real and design (as repository of a residual idealism) takes precedence over dwelling and occupation. I revisit this difficulty in Chapter Four, looking at new cities, and in Chapter Eight on an alternative Modernism. Here I dwell on the architectural metaphor in Descartes' text, drawing on a commentary by Claudia Brodsky Lacour (1996). I investigate a progressive purification of urban space from the seventeenth century onwards as a consequential abstraction of the city, purging it of dirt that denotes uncertainty; and compare the tension between thought and action inherent in Descartes' metaphor of drawing a line with that between imagination and displacement in literary Utopias. Finally, turning from abstraction to informed action, I note the act of digging up the land that marked a glimpse of Utopia in England in the 1640s.

Confidence and lack of certainty

Edward Soja reiterates that city air makes people free (Soja, 2000: 248). This is the saying inscribed on the gates of cities in the Hanseatic League, a trading network around the Baltic Sea. The phrase succinctly expresses the freedom of a merchant society from the ties of feudalism, and the free movement of trade across state boundaries. Initially this represented a shift from medieval to early modern society, and the progressive urbanization of Europe prior to its colonization of new worlds elsewhere. From the seventeenth century, this sense of a freedom to act independently is the concern of an emerging commercial class that drives a new rationalism in urban planning, as well as the agricultural, industrial, political and institutional revolutions of the eighteenth and nineteenth centuries. Together with the administrative class and professions such as medicine and law, the commercial class becomes the bourgeoisie, the first explicitly urban class whose class identity is integral to the concept of the city. Under this new class and its rational-commercial ethos, articulated in scientific knowledge, urban development in the eighteenth century produced planned city districts to replace the dark, twisting alleys of medieval city quarters. When, for instance, the old centre of Lisbon was destroyed by an earthquake in 1755, it was rebuilt as a commercial centre on a grid plan (Maxwell, 2002).

The main perceived threat was not of natural disasters but of disintegration of the social fabric. Crowded inner-city quarters were razed and replaced with broad vistas, illustrated by London's Regent Street and Trafalgar Square in the 1820s. In Paris in the 1850s and 1860s, Baron Haussmann drove new boulevards through the working-class quarters, constructing a bourgeois city of apartments and an environment designed to ensure order as well as deliver profits to speculators (Short, 1996: 176; Benjamin, 1999: 120–49). In the nineteenth century policing by other means, as it were, extended to urban improvement in the establishment of cultural institutions as well as digging sewers, while the poor were marginalized by being sent to geographical and/or social margins. I return to this in Chapter Four, and here simply note that a dual aspect to the growth of a commercially oriented Europe emerges: on the one hand rational space; on the other, fear of implosion. If the surfaces of modern European city life expressed confidence, then under the surfaces were personal and social insecurities that produced, parallel to the development of rational philosophy and urban planning, an at times hysterical repression of deviance and dirt – its perhaps necessary corollary. For example, Walter Benjamin juxtaposes the underground city of nineteenth-century Paris – its modern metro and ancient grottoes and catacombs – with the overground city of iron and glass arcades (Benjamin, 1999: 85). But that, of course, post-dated the intellectual acceptance of the model of an unconscious to which the repressed material of consciousness was displaced.

It is too easy to universalize such relations – as of appearance and exclusion – as the positive and negative sides of modernity; more interesting are the real and intricate histories of the modern period, from the colonization of the Americas through religious conflicts in the sixteenth and seventeenth centuries to the agricultural and industrial revolutions. These complex histories, producing changes in the structures of class and the European state, are the contexts of the new rationality. If the face of Reason seems serene, its body is the product of upheaval. It is, again, too easy to describe these upheavals as unprecedented (retrospectively applying the metaphor on which the chapter is based to its background material), yet it is difficult to imagine the scale of political insecurities and deathly impacts of military conflicts in this period, when wars of religion reduced the population of what is now Germany by perhaps as much as one-third. Stephen Toulmin writes, 'Across the whole of central Europe, from the mid-1620s to 1648, rival militias and military forces consisting largely of mercenaries fought to and fro, again and again, over the same disputed territorie' (Toulmin, 1990: 54). Of the 1630s, the period of the *Discourse*, he says, 'no one could see an end to warfare in Germany' (Toulmin, 1990: 55), adding that the search for a rational basis for knowledge was driven by a need to find an alternative to the religious conflicts taken as pretexts for wars. Peter Wagner, too, notes a 'situation of uncertainty that was experientially self-evident in a situation of devastating religious and political strife' (Wagner, 2001: 19). The search for rationality was, then, a search for freedom from chaos outside the contingencies of everyday conflict, on a scale that today might be seen as limited to the non-affluent world, or to a margin of the affluent world such as the former Yugoslavia in the 1990s. In the 1630s it happened in the middle of a Europe divided into what were still for the most part the dynastic territories of a ruling elite, where an emergent rationality occurred parallel to a growth of personal religious interpretation of texts increasingly available in vernacular languages.

For Wagner, Descartes and Hobbes were 'radical in their rejection of unfounded assumptions, but inflexible in their insistence on some definable minimum conditions of cognitive and political order that could . . . be universally established' (Wagner, 2001: 19). Toulmin reads Descartes' effort to find new certainties as an urgent reaction to the collapse of the political situation: 'for a generation whose central experience was the Thirty Years War, and a social destruction that had apparently become entirely out of hand, the . . . appeal of geometrical certainty . . . helped his program to carry a new conviction' (Toulmin, 1990: 62). But where to find certainty, except in the individual mind withdrawn from the world's destructiveness to the intimacies of contemplation? As it happens, the emergence of a unified self in early modern culture (Chapter One) brought its own insecurities as to how the self was to be recognized.

In a commercial society, identity was constructed through clothes, houses and the luxury goods available in cities. E. J. Hundert, writing on Bernard Mandeville's *The Fable of the Bees*, summarizes, 'These people were driven not merely to universal appetites for authority and esteem. Rather . . . outward displays of wealth were now widely accepted as an index of one's identity' (Hundert, 1997: 75). Hundert continues that commerce and sociability were related factors in the modern dynamics of a subject whose self-regard is based on that of others; and that Mandeville's insight was to see a new relation between motivation and action when social action is a pursuit of pleasure derived from the approval of others: 'Behaviour in public consisted of performances designed to win approval . . . whose success depended upon no genuine moral standard, but on how well social actors could satisfy their desires within established conventions' (Hundert, 1997: 76).

In the emerging art market of the Netherlands in the seventeenth century, the group portrait represented a vehicle for such public display. In a period of emerging subjectivity and increasing emphasis on the public performance of a self amid the fluctuations of a market economy, however, there were personal insecurities. Roger Smith writes:

> Whatever the subsequent modifications and intensifications of the sense of self by Romantic writers, by urban society or by modernist art or philosophy, it is possible in the late twentieth century to grasp and identify with the individualizing content of seventeenth-century expression. . . . [This] created possibilities for psychological experiences . . . There could, after all, be no psychology unless there is a psychological subject.
>
> (Smith, 1997: 50–1)

Caspar Barleus, who was probably present at the anatomy lesson of Dr Nicholas Tulp painted by Rembrandt, was a leading intellectual and, as Francis Barker writes, a 'noted neurotic, who wrote poetry in praise of Tulp's dissection . . . and dared not sit down for fear that his buttocks, which were made of glass, would shatter' (Barker, 1984: 115). Such an individualized neurosis depends on an individual subject, as on a vocabulary derived from the urban technologies of the time, such as glass blowing. According to Barker, Descartes, too, may have been present at Tulp's anatomy lessons, being himself an anatomist and living much of his adult life in the Netherlands.

Derivation of Descartes' architectural metaphor

Whether Descartes was there at the anatomy lesson is anecdotal, as is the story that he moved his lodgings 23 times during 20 years in the Netherlands, or that he had only one known sexual relationship (with a servant, producing an illegitimate child) and remained unmarried; or that his mother died when he was 13 months old and that, as a sickly child, he was allowed to stay late in bed at the Jesuit school to which he was sent at the age of 10 – a habit he retained throughout his life. Such data comprise the anecdotal material that is used in historical biographies, and in some psychoanalytic accounts of an individual's consciousness and dysfunctionality, but which Descartes sought precisely to eliminate in seeking certainty. So, when psychoanalyst Anthony Storr writes, 'Cartesian dualism . . . seems to have originated from the deprivations of his own early childhood' (Storr, 1976: 99), he may misread Descartes when he says his dualism is an almost total split of mind and body. More generally, I agree that Descartes' philosophy is grounded in uncertainty, and that it is a more extreme uncertainty than that of Michel Montaigne, expressed in his essays written in the 1580s as a personal, reflexive position, one element of which is that little can be known. Montaigne personalizes this in terms of a doubt as to what he can know, and Descartes extends this as a personalized proof of existence in the verification of a doubting subject. But Descartes also extends this into treatises on mathematics, geometry and optics which for him constitute his major work, the *Discourse* being more a contextual text in which he explains how he arrives at the position he adopts as objective author of those treatises.

What survives uncertainty is an uncertain self – the 'I' who speaks – extending to an idea of a consciousness which Descartes interchangeably calls soul or mind, and, as John Cottingham notes, to research into the physical operations of consciousness (Cottingham, 1997: 23). To return, though, to anecdotal evidence: Descartes met the mathematician Isaac Beeckman, who saw mathematics as a way to deal with problems in physics, in Breda in 1618. Descartes wrote to Beeckman saying he wanted to produce 'a completely new science which would provide a general solution of all possible equations' (cited in Cottingham, 1997: 9). Shortly after, taking military service as a means to travel, in the winter of 1619 Descartes was at Ulm in Germany on St Martin's Eve. In his home town, La Haye in France, this was a night for remembering the dead: 'Descartes suffered what some have interpreted as a nervous breakdown, while others . . . have construed it as the real start of his philosophical career' (Cottingham, 1997: 10). Fourteen years later, in Germany again during the Thirty Years War, returning from the Emperor's coronation, Descartes found himself alone and sat for the day in the warmth of a room heated by a tiled stove. Barker writes, 'meditating by the stove, considering strangely whether his body exists, [he] uses the wax that is to hand to prove that corporeal objects have no consistency or essentiality but extension in space' (Barker, 1984: 115). Descartes' solitude was habitual – he writes of his life in the Low Countries, 'Living . . . amidst this great mass of busy people . . . I have been able to lead a life as solitary and withdrawn as if I were in the most removed desert' (Cottingham *et al.*, 1985: 125–6, cited in Cottingham, 1997: 19). He observes that in his efforts to rid himself of uncertain knowledge he nonetheless learns from what in retrospect seems ill-founded, 'just as in pulling down an old house we usually keep the remnants for use in building a new one' (Cottingham, 1997: 18). Yet while much of the *Discourse* concerns the limits of knowledge from travel, books and the senses, all of which is anecdotal, there are passages in which the new is claimed as superior to the adapted (and by extension the abstract idea to the remembered scene). One way to read the *Discourse* is as a contest between anecdotal knowledge and abstract thought, or between the received and thereby reinterpreted and the newly calculated. It is as if the architect is faced with a pile of salvaged timbers and has to decide whether to use them or to commission new ones in the building of a house. Descartes does not resolve this tension, but in the following passage from the *Discourse* – which I quote at length because it seems a key passage, to be read as a whole and not broken up by my interpretation – uses an architectural metaphor to assert a radical claim for the new:

> I spent the day in a stove-heated room, with all the leisure in the world to occupy myself with my own thoughts. Among these, one of the first that came to my mind was that there is often less perfection in what has been put together bit by bit, and by different masters, than in the work of a single hand. Thus we see how a building, the construction of which has been undertaken and completed by a single architect, is usually superior in beauty and regularity to those that many have tried to restore by making use of old walls which had been built for other purposes. So, too, those old places which, beginning as villages, have developed . . . into great towns, are generally so ill-proportioned in comparison with those an engineer can design at will in an orderable fashion . . . And, if we reflect that nevertheless there have been at all times officials charged with the task of keeping watch over private building and making it serve the public interest, we will easily understand how difficult it is to achieve any degree of perfection by adding to the work of others. In the same way I fancied that half-savage nations, who had gradually become civilized, but who had made their laws by degrees . . . could never be as well regulated as those who, from the beginning of their associations, had observed the decrees of some prudent lawgiver.
>
> (Descartes, 1960: 44–5)

The appeal to newness can be compared to other passages in which a common-sense realism is regarded as superior to book-knowledge, and in which the claim to make new is ambiguous. There are elements of adaptation, too, in Descartes' text: Hassan Melehy cites Etienne Gilson's 1925 edition of the *Discourse* with its extensive notes on sources: 'most striking are the passages in which Montaigne's own wording may be seen, just slightly rewritten' (Melehy, 1997: 93). This rewriting includes the opening sentence of the *Discourse* (Melehy, 1997: 178, n.11). But if Descartes distrusts the accumulated knowledges of books in which the texts of previous writers are contained, and yet adapts them in his writing, is this a contradiction, or is it simply that Descartes adopts the literary form of the essay that Montaigne invented? I leave the question open.

To return to the architectural metaphor: old buildings or cities adapted through time are inferior to new ones designed by one person. In the style of Descartes' narration this is metaphor. Lacour gives a more engaging translation of one sentence than the version I cite above: 'the well-ordered towns and public squares that an engineer traces on a vacant plain according to his free imaginings [or fancy]' (Lacour, 1996: 33). This states and encapsulates the foundational metaphorical act of modernity: drawing a line. Lacour follows Gilson and gives the French: 'au prix de ces places régulières qu'un ingénieur trace à sa fantaisie dans une plaine' (Descartes, 1963: I: 579 in Lacours, 1996: 32). She describes this as 'the terms of a pseudo-opposition between the figurative book of the world and literal book learning' (Lacour, 1996: 33). Commenting on the term fantasy (*fantaisie*), she notes it is used prior to the Romantic opposition of imagination to reason, and that for Descartes it means thought 'unencumbered by material constraints' (Lacour, 1996: 33, n.4). I do not think it matters whether or not Descartes saw an early planned

28 *Histories and theories*

city such as Charleville (1605), to which such an image might refer; the image of an engineer drawing a line according to his imagination is an expression, after the thought, of the world as rationality. Nonetheless, the reference to public places denotes the same principle of regularity as the metaphor, and such regularity was evident in sites such as Place Royale and Place Dauphine in Paris, and the emergence of the model of a city plan. A history of early modern urban planning and design would show a development, though, beginning before Descartes' time and extending beyond it, of a transposition of accurate measurement – used in mapping a city for tax purposes as well as in visual representation – from description to prescription. Just as the history of the modern subject contributes, then, to understanding literary Utopias, so it is worth outlining here a history of urban representation as structured by the visual sense, and its metaphorical aspects, before returning to the meaning of Descartes' text.

Drawing the city

In the 1790s, the regularly proportioned vistas of Versailles, Karlsruhe and Potsdam informed Charles L'Enfant's plan for the new city, built on a swamp, of Washington DC (Sennett, 1995: 265). Richard Sennett sees the idea of a planned city as evidence of a desire to facilitate free circulation, on the organic model of the circulation of the blood published by William Harvey in *De mortu cordis* (1628). This fits the idea of cities as commercial sites requiring free movement of goods and citizens, and links to a later identification of the free circulation of bodies, water and air with health. Sennett writes, 'Harvey's revolution helped change the expectations and plans people made for the urban environment. [It] . . . led to new ideas about public health, and in the eighteenth century Enlightened planners applied these ideas to the city' (Sennett, 1995: 256). But the process is not simply one of cause and effect in which the ideas are causes and the practices are effects. The derivation of the process of drawing a city plan begins with Leon Battista Alberti's invention of a device for making perspectival drawings of a city's streets from a circuit of walls, presented as a standard procedure in *Descriptio Urbis Romae*. Like Descartes, Alberti emphasizes a common-sense dimension to his invention, a means that can be followed by anyone of average intelligence (Pickles, 2004: 128). Then, as the city plan has new uses including the measurement of urban plots for taxation (Pickles, 2004: 92–106), the standard type of the plan as if seen from the sky is normalized, in turn conditioning the way plans for new cities are drawn. But Alberti's device is not isolated from his formalization of perspective drawing (Damisch, 1995: 100–13; White, 1972: 121–6), nor from his affirmation of the gendering of domestic architecture (Wigley, 1992: 332–51). For Denis Cosgrove, the mode of terrestrial mapping 'marks the beginning of European Modernity' (Cosgrove, 1994: 271, cited in Pickles, 2004: 83). For Alberti, as for Descartes, the use of an objective viewpoint is ideological and gendered (Rose, 1993; Massey, 1994: 212–48), a cartographic gaze that objectifies what is seen through its lens. But for Descartes it is a gaze of certainty that no longer requires the corollary of measuring a real space, superseded by the abstract gesture of drawing a line that is, I think, more than mapping for practical purposes, but rather the beginning of modernity (see Venn, 2000: 39). From it follows the possibility to draw a society, and to engineer its social relations, its medium of exchange, its structures of power, the width of its streets and the length of the working day measured by public clocks – all this can be arranged in a regular and rationally proportioned way. Then are we free? Then can we draw a line under the constraining past?

Drawing a line

There is a tension between an idea of a city as a sequence of vistas in public spaces – from the eighteenth century drawn increasingly as a series of public spaces around which districts of the city coalesce – and the privacy of thinking subjects. Descartes declines the aim of reforming society, seeking only to reform his own thoughts – 'a foundation that is entirely my own' (*un fonds qui est tout à moi*) (quoted in Lacour, 1996: 35). He describes the book in which he says this as a story, or fable. As with More, there is a distancing here to avoid being drawn into religious controversies, and equally an invitation to refuse the authority of anecdote or the opinions of others. For Melehy, the first person narrative 'shows a desire to be autonomous . . . to institute the philosophical subject as purely autonomous, not dependent on a source' (Melehy, 1997: 100). This figures a non-situated subject in the absence of others whose perceptions may confirm the subject's observations, which, for John Dewey, constitutes a 'spectator theory of knowing' (Dewey, 1984: 195, cited in Wagner, 2001: 17). Wagner argues that this position is not derived from Plato but from Descartes' radical doubt: 'The spectator theory of knowing thus starts out from a radical scepticism . . . And it is in response to the doubts about the world around us . . . that the quest for certainty needs to proceed through a distancing from the world' (Wagner, 2001: 18). The space imagined as location for the distanced subject is ideally the blank ground on which Descartes' engineer draws a line – the subject's non-place, or Utopia, and the space projected onto the world for the purpose of its ordering. It expresses 'the symbolic brushing aside of all previously existing claims to know' (Venn, 2000: 107).

In the pre-modern period knowledge was a reinterpretation of past texts. The gesture of drawing a line 'at once erases this whole tradition . . . announcing the futurity of truth: knowledge is the yet unknown . . . awaiting proof and confirmation in the future' (Venn, 2000: 107). Couze Venn sees a real site of the *tabula rasa* in the New World, which is accurate in that the cities of New Spain were renewals of those of old Spain, but not in that, as Setha Low demonstrates, their plans were derived from the city forms of Mexico as much as from Spanish fortified cities themselves derived partly from the cities of the Moors (Low, 2000: 101–23). Yet New Spain was, like the West in the nineteenth century, a virgin land, and Venn argues that, as a blank space on maps, it could be effectively 'depopulated' by the device of dehumanizing its indigineous inhabitants. With extraordinary violence, the conquerors sought gold and the conversion of the heathens while categorizing them as Indians, sub-humans or savages (Venn, 2000: 111–21). This is instrumental rationality, however, the power-over of Bacon's scientific analysis, and not, I think, the rationality of Descartes' metaphor. It may be a fine line but it is there: between a metaphorical coming to terms with doubt and the inscription of a dominant society. Reading Venn's account of European modernity I am reminded of Zygmunt Bauman's argument that the Holocaust was not an aberration but the application of its rationalizing principles (Bauman, 1989). But as Primo Levi writes in *If This is a Man*:

> It is . . . necessary to be suspicious of those who seek to convince us with means other than reason . . . It is better to renounce revealed truths . . . to content oneself with other more modest and less exciting truths, those one acquires painfully, little by little and without short cuts, with study, discussion and reasoning, those than can be verified and demonstrated.
> (Levi, 1987: 396–7, in Griffin, 1995: 392)

Drawing a line reconsidered

Perhaps Descartes' metaphor of the engineer drawing a line suggests a brittle quality in abstract reasoning, while the image of the engineer cannot be disconnected from the warmth of the stove-heated room and the withdrawal it allows from a world of violent uncertainty. Looking back over his philosophical development and the knowledge it has made available to him, Descartes writes that 'Mathematics gave me most pleasure because of the certitude and evidential character of its reasonings; but I did not then perceive the true use to which mathematics can be put' (Descartes, 1960: 41). In the *Discourse*, mathematics and geometry offer certainty, and treatises on both subjects as well as optics were attached to the published text. Descartes writes of moving forward in his discursive efforts slowly, as if a person walking alone in darkness, 'with so much circumspection at every step, that, even if I made a very little progress, I should at least not fall' (Descartes, 1960: 48). His attempt to overcome doubt is then qualified: in his youth he studied 'geometry and algebra in mathematics' but 'when I examined them . . . besides the fact that both only deal with what is highly abstract . . . the former is so bound to the inspection of figures that it cannot exercise the understanding without greatly fatiguing the imagination' (Descartes, 1960: 49). Yet Descartes' rules of inquiry include the dismissal of what is merely experienced, and the orderly conduct of thought in which geometers reach 'their most difficult demonstrations' (Descartes, 1960: 50). From this he speculates that all kinds of knowledge are linked in a common method consisting of the removal of falsehoods. The certainty that remains is of the order of a thinking being existing: Descartes' *cogito*. So, the engineer draws a line outlining a space which exists as idea. The idea and the line are inseparable. The idea is known only in representation, and the abstraction of the line is inseparable from the text describing it. That is to say, while I read the ethos of the *Discourse* as utopian in seeking a world of certainty, I do not read it as prescription. Before moving to some of the implications of the search for certainty, I want to look at a detail in the text.

Descartes writes of an engineer (or architect – the terms are used interchangeably) drawing a line, making what we might now call a design. But Descartes has his own design. He writes: 'my design here is not to teach the method that everyone must follow . . . but only to show [*faire voir*] how I have tried to conduct my own' (Lacour, 1996: 34). In this he imitates architects who make designs of their own rather than copying others (see Lacour, 1996: 34, n.8). There is a nuance here, which cannot be translated into English, between design as plan and as intention. Lacour notes that the French *dessein* and *dessin* derive from the Italian *disegno* (drawing) and that the two forms became associated with different meanings, a mental intention, and a material plan. Descartes' design is to build a foundation for knowledge, to express which he employs the metaphor of a plan of regular proportion. Later in the *Discourse*, he writes of the use of a ground plan (*dessin*) in rebuilding a house (Lacour, 1996: 36). This indicates a transition from intent to plan, but in the passage quoted at length above Descartes dwells on the ambivalence of design. From this, it is possible to construct an argument that modern architecture and planning abandon that ambivalence, vesting an increasingly authorial and professional weight in the design of spaces. In contrast, a desire to purge falsehoods and a method of certainty remain in permanent tension in Descartes' text. In the tension between intention and plan is the act I take, following Lacour, as foundational for modernity: the drawing of a line. It is useful to quote Lacour here on its significance:

> The act of architectural drawing that Descartes describes is the outlining of a form that was not one before. That form would combine reason . . . with imaginative freedom. . . . It is not only new to the world, but intervenes in a space where nothing was, on a surface . . . where nothing else is. The order of its 'places regulières' is the image of imagination engineering a method that is free of historical and intellectual as well as physical constraints. Yet, what can it mean to have 'thought' such an image in one's own mind, rather than to have read, seen, or heard it elsewhere, in the course of conversation, which is to say, already *as* representation?
>
> (Lacour, 1996: 37)

Purge and erasure: urban cleansing and clearances

To draw a few threads together: the urge to make new is fraught with the tension between a distanced subject and an objectified ground of observation; but although, or because, the new is imagined in the conditions of the extant, a possibility of intervention remains, and is rational and reflexive. I dispute Lewis Mumford's rejection of rational thought when he writes of modernity as based on a scientific ideology he attributes to Descartes and Galileo, mythicized as 'the central component of the Invisible Machine, [which] reduced reality to the calculated, the measurable, the controllable' (Mumford, 1973: 21). Mumford ignores the fact that Descartes resists prescription, that his text is contingent on the conditions of its production. As Toulmin says,

> The 17th-century philosophers' 'Quest for Certainty' was no mere proposal to construct abstract and timeless intellectual schemas, dreamed up as objects

of pure, detached intellectual study. Instead, it was a timely response to a specific historical challenge – the political, social, and theological chaos embodied in the Thirty Years War.

(Toulmin, 1990: 70)

But remember Caspar Barleus at Dr Tulp's anatomy lesson: a bourgeois, liberal individual afraid to sit down. Lying behind incipient Enlightenment rationality are insecurities beyond its inherent philosophical or political tensions. These surface in the murderous plots of Jacobean drama. In cities, the need to deny insecurity increases in keeping with the requirement of bourgeois society to maintain calm, leading to removal of elements of the population whose presence fractures the image of a rational society. Steven Connor begins an essay on the modern subject by saying there are two stories about modernity: one about 'the apprehension of the self's autonomous self-grounding'; the other about the absence of God. Within the first story is 'the act of expelling all in authenticity and error from the self' (Connor, 1997: 203). The authentic is incompatible in this context with error; error must therefore be purged if rationality is to remain real. I lack the space to go into this but the model may be compared to a psychoanalytic model of splitting: that which threatens to shatter a mental unity or ideal image in the subject is projected onto others as a means of denial in the self, then repressed in rejection (as bad) of those others onto whom it is projected (see Klein, 1988: 262–89). Psychoanalysis privileges the individual over the social, but the process can be read in terms of the social, for example in discussion of spatial boundaries (Sibley, 1995: 2000).

For a society founded on rationality and productivity, and the pursuit of pleasure in the display of identity, the presence of dystopian elements denied society's realization of the ideal. Similarly, the non-productive – the vagrant whose insecurity was produced by social change, like the thieves in More's Utopia (dispossessed by land enclosures) – must be made invisible in a society embodying productivity. Hence the great purge, or erasure of the visible presence of the vagrant, classified together with the insane and bodily confined in institutional spaces constituting a state within the state (Foucault, 1967). After this first step, the miasma of the dead must be erased by removing burials to the periphery. Then the odours of living bodies must be deodorized, and their wastes removed to underground passages. Finally, the poor who especially emit odours as well as the insurgents who threaten the whole idealized edifice must be repressed. Ivan Illich writes on the commission appointed by the French parliament in 1737, for example, to investigate the effects of miasma produced by bodies in shallow graves in churchyards in the inner city, and the closure of the Cimitière des Innocents in 1780 (Illich, 1986) when it became a nocturnal meeting place for lovers. David Sibley, bringing the narrative into the twentieth century, describes geographical and cultural exclusion in media imagery as well as urban and suburban spaces (Sibley, 1995). Like a coin with head and tail, the two sides – idealization and expulsion of the non-ideal – are counterparts. Illich writes:

During the eighteenth century it became intolerable to let the dead contribute their aura to the city . . . The nineteenth century created a much more difficult task for deodorants. After removing the dead, a major effort was undertaken to deodorize the living by divesting them of their aura. This effort to deodorize utopian city space should be seen as one aspect of the architectural effort to 'clear' city space for the construction of a modern capital . . . a city without aura is literally a 'Nowhere', a *u-topia*.

(Illich, 1986: 52–3)

Illich and Sibley draw on Alain Corbin's *The Foul and The Fragrant* (1996). Corbin devotes a chapter to 'The Stench of the Poor', writing that after the search for polluting odours in cemeteries, hospitals, prisons, 'and all those sites where people confusedly crowded together to produce the undifferentiated odours of the putrid throng', a new project concerned 'the odours of poverty' (Corbin, 1996: 142). This penetrated domestic spaces once public spaces were cleared, and Illich notes the adoption of the water closet among the English upper classes in the 1800s, and cabinets in upper rooms still cleared by servants in France (Illich, 1986: 57). But, as he then points out, bodies, apart from the excrement they produce, give off odours which remain in clothing worn close to the body and in bed linen unless washed frequently, noting that the Convention decreed a right to sleep in one's own bed: 'to be surrounded by a buffer zone' (Illich, 1986: 58). The opportunity to deodorize themselves was open to the bourgeoisie, and denied to the poor, whom they saw as an undifferentiated mass. Corbin notes, 'The absence of intrusive odour enabled the individual to distinguish himself from the putrid masses, stinking like death, like sin, and at the same time to implicitly justify the treatment meted out to them' (Corbin, 1996: 143). With an impulse akin to that of Barleus, 'the ruling classes were obsessed with excretion . . . Its implacable recurrence haunted the imagination . . . [while] the bourgeois projected onto the poor what he was trying to repress in himself' (Corbin, 1996: 144). The urban type most subject to such projections, in Paris, is the rag-picker who collects carcasses and remnant matter from public highways to remove it to dumps on the city boundary for sorting into what is useful and can be sold, and what is not.

The rag-picker is not only one of the poorest of urban workers and 'the archetype of stench' (Corbin, 1996: 146), but transgresses boundaries between the city and its margin, between dirt and property, between living and dead matter. Corbin notes 17 reports on rag-pickers in 1822, and that 'whether it concerned excrement, prostitutes, or rag pickers, the fascination mixed with repulsion pervaded the discourse and governed the attitude of sanitary reformers and social researchers' (ibid.). The observing experts and the bourgeois class which employed them as public officials remained scrupulously separate from the poor they observed, producing myths to justify the division, for instance by saying the poor produced poverty themselves through idleness, alcoholism and sin. This convenience preserves the utopian image of the city, but also displaces cities as experiential sites to an abstract concept, like the medieval city on a hill that shelters people from wild nature and its uncertainties outside the city gates.

Modern Utopias

Strangely familiar ideas are carried over the border. If thought, rather than the natural world observed through the senses, is the site of regularity and certainty, a universality associated with nature as the pre-existing state of being is transposed to thought. Toulmin sees the search for certainty as bound into a quest for cosmopolis, a world of cohesion in nature and human society in which natural philosophy is reconstructed on geometrical foundations, everything 'underpinned in a brand new way' until 'From 1650, European thinkers were taken with this appetite for universal and timeless theories' (Toulmin, 1990: 83). Toulmin links this to the shift from feudalism to the emergent nation state with its bourgeois and constitutional structures. Permeating the process are reactions to the disintegration of Europe in the early seventeenth century, and a movement from received or religious truths to abstract reasoning. Reason is the means to realize freedom and the complete human happiness that denotes an end of the vicissitudes of history. In terms of the structure of a state, clearly no longer viable as a ruler's personal estate, the inception of rationality allows the possibility of a public realm in which empowered (which means bourgeois) citizens act as individuals in a collective society. Wagner writes:

> To arrive from the individual members at ways of constructing stable collectivities, an additional assumption about the guiding orientation or behaviour of the members needed to be introduced in the human sciences . . . the assumption is basically one of rationality . . . In liberal political theory, the capacity for rationality leads the individuals to enter into a social contract for their mutual benefit and to restrict the play of their passions to the private realm.
>
> (Wagner, 2001: 60)

Many of the difficulties of the rational comprehensive model of planning (Sandercock, 1998: 87–9) follow from a division of cities in zoning policies, and from the privileging of professional expertise as the location of rationality over the anecdotal knowledge of dwellers. Leonie Sandercock writes of this model, 'It offers the illusion of certainty and objectivity, and allays . . . "the Cartesian anxiety" – the fear that, if we reject the possibility of absolute truth . . . then anarchy is loosed upon the earth, and "anything goes"' (Sandercock, 1998: 89, citing Bernstein, 1985). So, in More's *Utopia*, Utopus, the founder of the state, bequeathed 'a complete plan of the city' (Choay, 1997: 145). It is immutable, like a model of a cosmos, certain and immune to human intervention except in ways that reproduce it in new forms of the same. But rationalism in the seventeenth century also had other, transgressive forms. It permeated practical Utopias proclaimed by the disempowered seeking to reclaim their rights to land after histories of enclosure which, as More observed via his character Hythloday (see Chapter One), made them thieves. I end the chapter, then, with a very brief description of the radical movements of the English Revolution, not as an antidote to Cartesian rationality but as a parallel articulation of that rationality, equally contingent on the conditions of its production.

Radical settlement in the English Revolution

The radical groups of the 1640s, such as the Diggers, Levellers, Ranters and Quakers, were deeply religious, believing that the second coming of Christ was at hand after the upheaval of society in the war between King and Parliament. They enacted the millenarian New Age in the present as transformation of society and individual, and in the midst of an upside-down world. Christopher Hill notes recurrent bad harvests, rising food prices, declining money wages, higher taxes, new excise duties, a disruption of landlord–tenant relations when church lands were sold to individual landlords, and a return of soldiers to an already depressed rural economy after the wars between King and Parliament. Gatherings of poor people on the roads seized merchants' corn. A tract, *The Mournfull Cries of Many Thousand Poore Tradesmen*, stated 'Necessity dissolves all laws and government, and hunger will break through stone walls' (Hill, 1975: 108). These are conditions in which the poor cease to have anything to lose. Norman Kohn says of the Ranters, 'Many felt the time had come when God was pouring out his spirit upon all flesh. Ecstasies were everyday occurrences, prophecies were uttered on all hands' (Kohn, 1970: 288). Those who sought to go beyond the relative freedoms of the Commonwealth – representing the interests of the urban commercial class and landed gentry, from whose ranks Parliament was elected – made no significant division between political and social, or religious visions. Gerrard Winstanley wrote in 'The Earth Shall Be Made a Common Treasury' (1649), 'And I had a voice within me bade me declare it all abroad . . . there shall be no buying or selling, no fairs or markets . . . There shall be none lords over others, but everyone shall be a lord of himself' (in Carey, 1999: 67). It is an immanent, revelatory revolution: office is abolished and property held in common as enactment of the life to come. For the Diggers (called the True Levellers) God was the 'Head Leveller' (Hill, 1975: 132). But integral to these currents portrayed, as above, as religious radicalism, is the rationality that is equally the product of seventeenth-century society. When Winstanley wrote that 'Freedom is the man that will turn the world upside down, therefore no wonder he hath enemies' (Coates, 2001: 22), he expressed both religious feeling and rational thought, in a vision of a radically reorganized society that would enact the principle of freedom on which rationality is based.

The Diggers' programme included communal ownership of land and goods, the abolition of money, free universal education, replacement of the legal system and community service as punishment for less serious crimes. In April 1649 a group of Diggers dug up wasteland at St George's Hill, Surrey to plant vegetables (Hill, 1975: 107–13). Up to two-thirds of England was wasteland, and food was in short supply. The Diggers thus offered a viable solution to rural poverty (Carey, 1999: 68). David Petegorsky writes:

> Society has become aware of the fact that the world now affords limitless potentialities of development and exploitation . . . the middle classes increasingly perceive that they alone wield the key to its entrance; for they alone could adequately exploit the opportunities that presented themselves.

> New enterprises demanded investment on which no return could be expected and which none but they was prepared to venture. New machines . . . capital outlay which only they could supply. . . . Above all, the new age demanded initiative and imagination which were to be found only in those classes that were not shackled to the past.
>
> (Petegorsky, 1999: 16–17)

And for David Pepper the Diggers are the 'communists of the 1650s', known for 'their beliefs and actions supporting land ownership as everyone's fundamental right, and the idea that the land's fertility will be improved by working it with love and a communal spirit' (Pepper, 1991: 26). Hill emphasizes the economic aspect of Winstanley's vision and its impact on agricultural organization:

> Winstanley had arrived at the one possible democratic solution . . . The economic arguments . . . were overwhelming. England's growing population could be fed only by more intensive cultivation, by bringing marginal land under the plough. . . . Collective cultivation of the waste by the poor could have had the advantages of large-scale cultivation, planned development, use of fertilizers, etc. It could have fed the expanding English population without disrupting the traditional way of life . . . The Diggers sowed their land with carrots, parsnips and beans – crops of the sort which were to transform English agriculture in the seventeenth century by making it possible to keep cattle alive throughout the winter in order to fertilize the land.
>
> (Hill, 1975: 130)

It might not be fanciful, then, to see the Diggers as an intentional community for whom the total transformation of society both prefigured the post-apocalyptic age *and* enacted it in advance, bringing the age of the spirit of medieval millenarianism (see Bloch, 1986: 509–15) into present personal and collective consciousness. The integration of the personal and the collective (or political) is paralleled in an integration of religion and rationality, in which pamphleteers analysed economic problems scientifically while retaining their religiosity, and in a religious tolerance translated as a demand for social egalitarianism. Then claims to social rights are expressed as return to natural rights and as renewal of historical precedent. Hill (1975: 87) notes the Levellers claimed their historical rights as Englishmen and a natural right to liberty, while the Diggers claimed communal ownership of land as a natural right, using innovative means to cultivate it.

In the 1640s, divine insight replaced the divinely given power of kings but did so as a rational outcome of a claim by the poor to subject status. The certainty of belief that drove religious parties to murder each other in the wars of the 1630s in Europe, and the certainty of abstract reasoning, express profound uncertainty, however. To draw a line is one response, turning away from the mess of history. The act of digging up St George's Hill is another, in the certain knowledge that heaven's kingdom will replace the corrupt state. The promise of modernity is in both cases a radical transition.

3 Planning harmony
Charles Fourier and utopian socialism

Introduction

In the previous chapters I examined literary and metaphorical utopian thought in the early modern period in Europe. Thomas More and Tommasso Campanella were critical of conditions in their own societies via the device of a narrative describing an imagined, far-away elsewhere. No explicit link was made to change in the writer's own society, and the fictional world described was not intended as a plan, being merely a traveller's tale. In a related but not identical way, Descartes intends to share a personal reflection but has no design for a new society. Writing in an introspective space established by Montaigne in the 1580s, Descartes seeks certainty because he doubts, but finds that the only factor of which he can be certain is the mental life of a doubting subject, from which he moves to the certainty of self-contained numerical and geometric systems.

In this chapter I move to the nineteenth century, and to utopian writing envisioning a practical transformation of society – a plan no longer self-contained but applicable to the world in which the writer lived. I focus on the work of François Marie Charles Fourier (1772–1837) because his writing gives a comprehensive picture of what a utopian society might be like, framed by the French Revolution of the 1790s in which (as in England in the 1640s) the world was turned upside down. Fourier's writing may seem encapsulated in the histories that informed it; his model of a phalanstery – a settlement of 1,680 people of carefully balanced but diverse psychological types – and his idea of a libidinization of work may seem far-fetched now: but they raise issues that reappear, for example, in Herbert Marcuse's effort to fuse Marxism and psychoanalysis in *Eros and Civilization* (1956).

I begin by outlining the context in which Fourier developed his theory, then look at his ideal of social harmony, the phalanstery, and his concern with human passions as a basis for social organization. Finally, I compare Fourier's ideas briefly to those of Robert Owen, and cite the case of the Charterville Allotments as a practical demonstration of the communality that characterizes much nineteenth-century utopianism.

Fourier – background

Fourier was born in Besançon in 1772, the son of a prosperous cloth merchant. He disliked his father's bourgeois world, spending much of his childhood in retreat from it, studying a collection of atlases in his room, and raising a collection of plants: 'he turned his room into a veritable greenhouse. Pots and flowers were carefully arranged by colour, size, and species in symmetrical patterns which he allowed no one to disturb' (Beecher and Bienvenu, 1983: 4). Despite avowals that he would never work in commerce, in 1791 he was apprenticed to a merchant in Lyon. Then, because the terms of his father's estate were that he could draw on the major part of his inheritance only if pursuing a career in trade, in 1793 he set up his own business in colonial goods. Lyon was a large enough city to have a critical mass of buyers for such goods but in 1793 the city revolted against the Convention, and Fourier's goods were requisitioned. He was forced to flee, and after a period of conscripted labour took up employment again in the textile trade.

The rest of his life was spent in orderly obscurity in a succession of rented rooms that he changed regularly. He was not a bohemian or a dandy but in his writing mixed satirical criticism of the bourgeoisie with a detailed elaboration of a new society based on his discovery, which he saw as scientific, of the passional basis of human destiny.

Fourier's utopianism is, in retrospect, at odds with the scientific socialism evolved by Karl Marx and Friedrich Engels, in that he foresees the coexistence of classes rather than class struggle, but Engels writes of him sympathetically:

> Fourier is not only a critic; his imperturbably serene nature makes him a satirist, and assuredly one of the greatest satirists of all time. He depicts, with equal power and charm, the swindling speculations that blossomed out upon the downfall of the Revolution, and the shopkeeping spirit prevalent in, and characteristic of, French commerce at the time. Still more masterly is his criticism of the bourgeois form of the relations between the sexes, and the position of woman in bourgeois society. He was the first to declare that in any given society the degree of woman's emancipation is the natural measure of the general emancipation.
>
> (Engels, 1908: 64–5)

During the Directory (1795–9), a period of free-market economics and fluctuations in prices after the Jacobin attempt at a planned economy, Fourier became a travelling salesman before returning to work for his original employer in Lyon in 1800. Jonathan Beecher and Richard Bienvenu comment:

> Condemned to a career in the lower reaches of commerce, Fourier was too precise in his habits ever to fall into dire poverty. He always managed to earn just enough to permit himself a few indulgences, and he kept up appearances as best he could. But for the next thirty-seven years, while he mapped out the contours of his ideal city and chronicled the glorious adventures and amorous

intrigues of its inhabitants, he was to lead a solitary life of dull work and trivial distractions.

(Beecher and Bienvenu, 1983: 10)

Until his death in 1837, Fourier developed a critique of the values and practices of the bourgeoisie whose closed, conservative lives he saw in his childhood. He saw, too, the poverty of a proletariat whom the employers exploited. In 'The Vices of Commerce', he satirizes bourgeois morality in terms of a not infrequent resort to bankruptcy as a means to defray debt, writing of smuggling, usury, speculation, hoarding and cheating as the similarly corrupt practices endemic in bourgeois life (see Beecher and Bienvenu, 1983: 116–18). Fourier drew up a table of types of bankruptcy, in three orders, nine types and 36 species: 'This list could easily be tripled or quadrupled. For bankruptcy has become such an art that every day someone invents a new species' (in Beecher and Bienvenu, 1983: 118). Beecher and Bienvenu comment, 'If the Lyon insurrection disgusted Fourier with revolutionary politics, it was the financial chaos of the Directory that shaped his economic views' (Beecher and Bienvenu, 1983: 6). But perhaps more interesting is the systematic way Fourier reacts to those circumstances, always listing types and sub-types, recapitulating a model of hierarchy just as he created a horticultural orderliness in his childhood collection of plants. The same applies in his theory of labour: people do the work to which their temperament prepares them, in the company of others to whom they are attracted according to an intricate categorization of types. Fourier writes:

> By working in very short sessions . . . every member of Harmony can perform seven or eight different kinds of attractive work in a single day. . . . This method . . . impels men and women to flit from pleasure to pleasure, to avoid the excesses that ceaselessly plague the people of civilization.
>
> (Beecher and Bienvenu, 1983: 275–6)

Richard Francis summarizes, 'we go about our business as if participating in a masquerade . . . But in this context at least masquerade is synonymous with utopia because our whims prove to be structured, our frolics architectural' (Francis, 1997: 81).

Fourier's critique is derived from his experiences in the late 1790s, when he saw at first hand how the commercial class profiteered from manufactured shortages of goods including food, from speculation and from the introduction of paper money. He may have read utopian tracts such as the pamphlets of François-Joseph L'Ange (circulated in Lyon), but his idea of a society based on passional types (as I explain below) appears to draw also on the emerging science of statistics and on a general scientific rationalism. Yet, like More, the intricate detail of Fourier's texts lends them plausibility, a realism at the level of detail masking a fantasy perhaps at the level of the scheme as a whole. This may in turn reflect the fact that Fourier's desire for a life of creative curiosity was contradicted by the necessity of working in the cloth trade of his father. I wonder if his theoretical enterprise is a reaction against the circumstances implicit in his remark, 'they got me into commerce against my will. I was lured to Lyon by the prospect of a trip . . . I deserted, announcing that I would never be a merchant' (Beecher and Bienvenu, 1983: 107).

Fourier's travels as a salesman took him to other countries where the conditions of textile workers were as degrading as those in the Lyon silk trades. In 1812 his mother died, leaving him a legacy that enabled him to take up writing full-time for five years from 1815, and he moved to Belley, a country town near Lyon, to live with relatives. In 1821 he returned to Besançon and found a publisher for his first major work, *Traité de l'association domestique-agricole*. He sent 1,000 copies to Paris, going there himself in 1822 to promote it. His efforts to interest government ministers, officials, bankers, scientists, philanthropists, journalists and academicians bore little fruit, and his money ran out. Fourier went on strike, refusing to write as redress for the world's refusal to recognize his work. He returned to Lyon in 1825, resuming the quiet, orderly life of a clerk, keeping cats, growing plants, and in his neat dress and regular habits seeming conformist. In 1837, still in lodgings, he was found dressed in a black frock coat, lying dead by his bed. The inscription on his grave says, 'the Series distribute the harmonies. The Attractions stand in relation to the destinies' (Francis, 1997: 77).

Fourier's theories

Fourier's theory rests on his identification of a spectrum of differential temperament, from generic to specific types. Work is a varied social activity, a pleasurable expression of the whole self whereby people work with those to whom they are attracted, in ordered variety. His theory of 'the destinies of all the globes and their inhabitants' (Beecher and Bienvenu, 1983: 31) was first published in *Bulletin de Lyon* (3 December 1803, or 11 Frimaire XII in the revolutionary calendar). Of 16 successive kinds of society, 'only three are to be seen on our globe: Savagery, Barbarism and Civilization. Soon they will come to an end, and all the nations of the earth will enter the fifteenth stage which is Simple Harmony' (in Beecher and Bienvenu, 1983: 81–2). Savagery, barbarianism and civilization are but rungs on the ascent to Harmony, while other times usher in 'perpetual peace, universal unity, the liberty of women' (ibid.). Appealing to Napoleon to become the founder of Harmony and liberator of the world, Fourier admits some readers will see him as a dreamer. To them he says:

> Blind savants, just look at your cities paved with beggars, your citizens struggling against hunger, your battlefields and all your social infamies. Do you still believe that civilization is the destiny of the human race? Or was J.-J. Rousseau right in saying of the civilized: '. . . there is a disorder in things, the cause of which we have not yet fathomed.'
> (in Beecher and Bienvenu, 1983: 81–2)

Fourier references a fundamental ambivalence in French philosophy in which progress is both beneficial and harmful. From Rousseau, Fourier claims a projected original state of human nature on which to base a further claim to rights which once belonged to all, and can be reclaimed. This follows pre-revolutionary political expression as the redress of diversions from a naturally right path when kings are misled by bad ministers, so that revolution re-establishes what once was. This mode of argument was used during the Revolution, as when François-Noel Babeuf argued at his trial:

> Man's condition ought not to have deteriorated in passing from a state of nature to a state of social organization. In the beginning the soil belonged to none, its fruits to all . . . if private ownership of public wealth is only the result of certain institutions that violate fundamental human rights; then it follows that this private ownership is a usurpation; and . . . that all that a man takes of the land and its fruits beyond what is necessary . . . is theft.
> (Carey, 1999: 175)

Babeuf adds, 'Society must be made to operate in such a way that it eradicates . . . the desire . . . to become richer, or wiser, or more powerful than others' (in Carey, 1999: 177). This puts faith in equality as a natural condition, and echoes Abbé Morelly's call for an elimination of greed in *Code de la Nature*, first published in 1755. Morelly's argument is, as Gregory Claeys glosses, 'linked to millenarianism, as well as to ideas of primitive apostolic Christianity, paradise, heaven, and the Garden of Eden' (Claeys, 1994: xviii). Babeuf replaces this with a secular egalitarianism that goes to the root of discrimination in the concept of equal ability but states his case in terms of a return to a natural state of society. Fourier's formative years were the 1790s and he, too, adopts a concept of natural rights. He combines it, looking to another strand of French philosophy, with a scientific rationalism stated in a claim that his theory is based on his discoveries.

Returning to Fourier's writing, key to his project was the liberation of the human passions from the deceits of bourgeois marriage and the limitations of bourgeois, competitive society. I use the translation *passional* attraction (rather than passionate) for the mutual attraction of people of complementary personality types he describes. He writes about orgies in the phalanstery but his theory is more about making work fulfil human needs (though I do not want to devalue the erotic content of Fourier's theory, and return to it below). Work, then, is pleasurable in acts of cultivation and propagation, manufacture or husbandry, which are a vehicle for social mixing. Underpinning passional attraction is the need of types to find others who complement themselves, set out in precise detail. Frank Manuel writes, 'provision is made for the specialized psychic needs and desires of each stage of the life-cycle . . . Fourier's utopia operates without any concept of repression' (Manuel, 1973: 83).

This begins as an effort towards a more effective form of agricultural production, in *Théorie des quatre mouvements* (1808):

> More than once people have supposed that incalculable savings and ameliorations would result if one could bring together the inhabitants of a village in an

industrial society, if one could associate two or three hundred families of unequal wealth according to their capital and their work. At first the idea seems completely impracticable because of the obstacle that would be presented by the human passions . . . at least eight hundred [people] are necessary to establish a NATURAL or ATTRACTIVE association. I mean by these words a society whose members would be inspired to work by rivalry, self-esteem and other stimuli compatible with self-interest.

(Beecher and Bienvenu, 1983: 96–7)

This is the theory Fourier extends in later texts as a diversity systematized as a series of passional types, requiring a certain critical mass to provide all possible effective (or elective) affinities. This group of 1,680 people is the Phalanx, and inhabits the specially designed environment of the phalanstery. Ultimately, a large number of phalansteries will replace cities, evenly distributed through France to usher in the world of Harmony.

Fourier writes, in his 1808 text, that an agricultural association of 1,000 people has the advantages of common use of assets such as a granary, wine-vat and ovens. Members are treated equally despite having different backgrounds while 'The strongest passion of both peasants and city people is the love of profit' (Beecher and Bienvenu, 1983: 96–7). Hence each person acts according to received ability and extant attraction to others, and all seek gains that can be made only by cooperation. Fourier insists, 'Each worker must be an associate who is compensated by dividend and not by wages' (in Beecher and Bienvenu, 1983: 274). Men and women do several types of work in a day, in amicable groups, in elegant and clean surroundings, and receive an income adequate to well-being: 'This guarantee must free them from all anxiety either for their own welfare or that of their dependents' (in Beecher and Bienvenu, 1983: 275). In his later writing Fourier goes into more detail, identifying pear cultivation within fruit growing, then seven types of pear from the hard to the juicy to the soft-sweet, and a series of tasks in cultivating each type of pear, for example, to produce elaborate tables. This also applies to sex in the phalanstery: 'all men have a penchant for polygamy. The behaviour of civilized ladies shows they have the same penchant' (in Beecher and Bienvenu, 1983: 334). Fourier proposes a sexual minimum like a minimum wage: 'society should grant a minimum of satisfaction to the two senses of taste and touch . . . The body may suffer if they are not gratified' (in Beecher and Bienvenu, 1983: 337). Then he says:

We are going to discuss a new amorous order in which sentiment . . . will enjoy an unparalleled prestige and will endow all social relations with a unique charm. How will sentimental love maintain this dominion? Through the fact that physical impulses . . . will be fully satisfied. Through the fact that the need for physical gratification will no longer be regarded as any more indecent than the appetites of the other senses . . . Only by satisfying the need for physical love, will it become possible to guarantee the development of the noble element in love.

(in Beecher and Bienvenu, 1983: 340)

The phalanstery

In 'The Establishment of a Trial Phalanx', Fourier sets out the material conditions required for a pilot Harmony, in case a philanthropist or ruler will establish it:

> An association of 1500 or 1600 people requires a site comprising at least one square league of land . . .
>
> A good stream of water should be available; the land should be hilly and suitable for a variety of crops; there should be a forest nearby; and the site should be fairly near a large city but far enough away to avoid unwelcome visitors.
>
> The trial Phalanx will stand alone and it will get no help from neighbouring Phalanxes. As a result of this isolation, there will be so many gaps in attraction, so many passional calms to fear in its manoeuvres, that it will be particularly important to provide it with the help of a good site for a variety of functions. Flat country, like that surrounding Anvers, Leipzig or Orleans would be quite inappropriate . . . It will therefore be necessary to select a diversified region, like that near Lausanne, or at the very least a fine valley provided with a stream and a forest, like the valley from Brissels to Halle. A fine location near Paris would be the stretch of land between Poissy and Conflans, Poissy and Meulan.
>
> (Beecher and Bienvenu, 1983: 235)

At Belley, in 1818, Fourier writes that phalansteries are to replace the cities where a concentration of wealth has delivered venality and corruption, but he sees the countryside as no better: 'secret orgies as well organized as in a big city' take place, and girls are deflowered at 10 with the compliance of their parents 'as dispassionately as the mothers of Tahiti collaborated in the prostitution of their daughters' (in Beecher and Bienvenu, 1983: 169–70). The phalanstery is neither urban nor rural, like a small city in a rural setting, or a large monastery or country house. It should be as perfect and unified as the terrain allows. In its centre

> should be a place for quiet activity; it should include dining rooms, the exchange, meeting rooms, library, studies, etc. This central section includes the temple, the tower, the telegraph, the coops for carrier pigeons, the ceremonial chimes, the observatory, and a winter courtyard adorned with resinous plants. The parade grounds are located just behind the central room.
>
> (in Beecher and Bienvenu, 1983: 240)

In one wing are workshops, in the other ballrooms and meeting rooms where visitors are entertained, as well as committee rooms. Outside are stables and farm areas, and pleasant gardens. Children are housed on a mezzanine. The interior connecting spaces are like the galleried arcades of Paris that Fourier saw in the 1820s, when they housed shops selling luxury and colonial goods like those in which he once traded. Everyone uses the arcades, and Fourier writes that they will make palaces seem like places of exile:

All the portions of the central edifice can be traversed by means of a wide gallery which runs along the second floor of the whole building. At each extremity . . . are elevated passages, supported by columns, and also attractive underground passages which connect all the parts of the Phalanx and adjoining buildings. Thus everything is linked by a series of passage-ways which are sheltered, elegant, and comfortable in winter thanks to the help of heaters and ventilators.

(in Beecher and Bienvenu, 1983: 243–4)

In Paris the arcades were refuges from rain and carriages hurtling through narrow streets. The shops were at ground level, the upper floors housing workrooms used also for sexual transactions (described by Emile Zola in *Thérèse Raquin* – see Benjamin, 1999: 875). Carriage entrances at ground level in the phalanstery mean the arcades must be at the second floor, but there are passages under the building, too – an image of mental and bodily movements, idea-paths and bowels, or the mechanism of production and waste in manufacturing, perhaps.

Benjamin reads the phalanstery as a machine, arguing that a collective consciousness is sparked by appearance of the new but reverts to a primal past of, as it were, collective utopian memory:

These relationships became discernible in the Utopia devised by Fourier. Their innermost origin lay in the appearance of machines. . . . The phalanstery was to lead men back into relations in which morality would become superfluous. Its highly complicated organization resembled machinery. The imbrications of the *passions*, the intricate combination of the *passions mécanistes* with the *passion cabaliste*, were primitive analogies based on the machine, formed in the material of psychology. This machinery, formed of men, produced the land of Cockaigne, the primal wish-symbol, that Fourier's Utopia had filled with new life.

(Benjamin, 1997: 159–60)

For this, a critical mass of passional types is vital. It is not necessary to list all 810 of them arranged in series and sub-series, but the classification begins from 12 basic types of passion which individuals carry in differing combinations; when all potential relations are made between these categories, the result is Harmony.

The passions

If the phalanstery is the built form of a life of unity, in which joy is unrepressed, then its social form is a settlement in which life is organized according to passional attraction. Passions operate continuously, immune to reason or reflection. In Freudian terms they would be the drives of the Id (*Es*) prior to organization in the Ego (*Ich*). Fourier sees the dominant organization of society as based on the repression of drives and sublimation of a desire for instant gratification, diverting and corrupting human feeling. By allowing intercourse between people with mutually attractive passional natures, society becomes non-repressive. Manuel summarizes, 'The bringing together of . . . psychological types under one roof was a precondition for happiness . . . otherwise the variety of relationships necessary for total self-fulfilment . . . would be lacking' (Manuel, 1973: 83). The critical mass of the phalanstery's 1,680 dwellers therefore includes all passional types.

Barbara Goodwin writes that for Fourier, the ills of civilization 'are essentially caused by the subversion of human nature' (Goodwin, 1978: 15), and can hence be remedied in a liberation of passional attraction. From this it follows that the libidinal life of a society is integral to everything. The difficulty is that while Fourier appeals to an apparently natural capacity for joy, the organization of the phalanstery, while differing from that of the dominant society, is no less unnatural as a product of human social organization, breaking with past precedent, than the model to which it is presented as the alternative. Fourier appeals to history, citing the Spartans as having loaned their wives to all comers, while French Romanticism represented far-away places in an imagined Orient as sites of sexual corruption; but the selectivity of both histories is in effect a vehicle for projection. Nonetheless, while Orientalism displaces a repressed sexuality to a far-away, as though inverse utopian, site, Fourier seeks to reintroduce the sexual freedom that is denied in that repression, at least removing the cause of desire's corruption as licentiousness.

Fourier sets out a series of amorous types in the phalanstery. Among youth, vestals postpone sex until they are 18; damsels experience it sooner, but anyone may join or leave either group at will. For those who choose, a state of amorous nobility in later life sets cerebral passions above bodily gratification. This series has seven categories, merging the active and passive with degrees of the low or saintly. Francis compares the series to the mathematical Mandlebrot set, describing Fourier as 'extrapolating from physical reality to its abstract formal principles' (Francis, 1997: 76). Fourier categorizes the passions into sets of five, four and three: the first five types denote the five senses; the second four denote four kinds of attraction – friendship, love, ambition and familial sympathy; the last three are the mechanizing passions – the cabalist type who joins intrigues, the butterfly type seeking variety of experience, and the composite type seeking physical and spiritual pleasures. Each type has a range of inflections. Each man or woman has a passional constitution dominated by one, two, three or more passional categories. Beecher and Bienvenu comment, 'the passions were benign and harmonious if afforded maximal expression. The source of evil was the unnatural set of institutions and moral codes known as civilization' (Beecher and Bienvenu, 1983: 41).

Fourier describes a variety of erotic adventures in Harmony. Sex is like the pleasure of taste, to be varied and openly enjoyed: 'In Harmony social relations are organized in such a way that no amorous intrigue can remain unknown' (Beecher and Bienvenu, 1983: 365). The class of abstaining, amorous nobility is open to anyone setting friendship over sexual love, but most inhabitants of Harmony opt for free love. In an exotic fantasy set in a fictional near-East, ironically similarly to the scenarios of Orientalist art, Fourier describes the experiences of a band of Harmonist adventurers captured by bacchantes: 'When the Head Fairy waves her wand a semi-bacchanalia gets underway. The members of both groups rush into

46 *Histories and theories*

each other's arms, and in the ensuing scramble caresses are liberally given and received' (in Beecher and Bienvenu, 1983: 389). He does not describe sex acts, and had little vocabulary with which to do so from the experience of his own life. Manuel dismisses Fourier as a 'pathetic little bourgeois salesman' (Manuel, 1973: 83) implying that Fourier's vision of a sexual minimum is wishful thinking. But I refuse to dismiss the erotic in Fourier's writing. The text from which the above extract is taken is both quaintly coy and dulling at the same time in its highly organized, repeated erotic encounters, but this is the problem of over-specification again, not the content of a life of joy. That joy, or *bonheur*, is evident in Baudelaire's poem '*L'invitation au voyage*' which Marcuse cites in an essay written in the 1940s on contemporary French literature, in which he argues that a literature of intimacy is the site of freedom in times of total oppression (Marcuse, 1998: 199–214). It resonates in an informal way, from the 1960s, with a description of the San Francisco Diggers as worker-priests of the counter-culture, who 'claimed that "you don't say love you do it", and . . . did it in the streets, in the road, in the park' (Farrell, 1997: 219; see also Chapter Five below). Zentrum für Experimentelle GesellschaftsGestaltung (Centre for Experimental Culture Design) in Germany also works with love and sexuality in a community practising free love (see Case Nine). Beecher and Bienvenu see Fourier as 'not a critic of his society but an enemy. He always insisted . . . his attack on civilization was radical, absolute, uncompromising' (Beecher and Bienvenu, 1983: 23). Fourier did attack the class into which he was born and by which he was employed, and became bitter at rejection of his ideas. But to me the tone of his writing expresses sympathy for human subjects who might yet be happy even if it is undermined by a level of detail that becomes prescriptive – like More's *Utopia* but as plan instead of traveller's tale – while Fourier fuses an almost alchemical tabulation with scientific rationalism. Francis writes of the grave inscription cited on p. 40 that the first sentence describes 'the macrocosmic order' and the second 'focuses on the microcosm, asserting that human energies . . . are directed at teleological fulfilment' (Francis, 1997: 77). Perhaps the problem is teleology, the idea that history can be planned. Manuel notes that nineteenth-century Utopias 'are organizmic and hierarchical, rooted in "scientific" theories of biology and physiology' (Manuel, 1973: 82). But I wonder if it is helpful to read Fourier as if he wrote fiction, or whether the novel is a better vehicle for this kind of vision. To test this, I move now to the opposite polarity of practical Utopias in the nineteenth century, before resuming discussion of Fourier and utopian socialism in comparison.

Owenite communities

Among intentional communities of the nineteenth century are the Owenite settlements of New Lanark, Harmony Hall and Concordium (see Coates, 2001: 73–4, 94–6, 166). This conflates Owen's experimental, industrial settlement at New Lanark with those later settlements in which his followers extended his vision, as in Harmony Hall in Hampshire, from 1839. New Lanark and Harmony Hall were both built utopian settlements based on cooperative values, nonetheless, and resemble the phalanstery. Both Owen and Fourier plan their ideal worlds in detail, Owen seeing an ideal community as consisting of 800 to 1,200 people undertaking agricultural and industrial work, with free education. Stedman Whitwell's design for an Owenite community (Coates, 2001: 75; Lockwood, 1905, opp. p. 70) resembles an image of the phalanstery by Victor Considérant (Sadler, 1999: 119, fig. 3.6) used in the Fourierist newspaper *L'Avenir*. George Lockwood, writing in 1905 on the Rappite community at New Harmony in Indiana, sees Owen and Fourier as both 'theoretico-practical' social architects fusing 'enunciation of social theories with actual experiments' (Lockwood, 1905: 2). In the course of asserting the greater importance of New Harmony over the utopian community of Brook Farm (see Francis, 1997: 35–139), Lockwood says, 'Owenism was the forerunner of Fourierism' while Owen declared that Fourier 'learned all he knew of communism from a study of Owenism' (Lockwood, 1905: 3). I am not in a position to know if Fourier borrowed his ideas from Owen. One approach would be to examine carefully the dates of first publication of each writer's texts, and the extent to which either's work was available (in translation or the original) in Britain, France and North America. As a theorist rather than a historian I take another approach, asking to what extent the ideas articulated by Owen and Fourier are more than seemingly or incidentally similar. Taking the images of ideal settlements, then, both have an outer rectangle of three-storey buildings and an inner open space in designs resembling either a military headquarters with parade grounds, a monastic settlement with enclosed gardens, or a palace or country house of a formal design. Whitwell uses more gables. Considérant adds wings on either side projecting forward of the central square. Both use gated towers to separate the inside of the settlement from the outside world. But these are not the most meaningful dimensions of the communities envisaged compared to the work undertaken within them (and neither Owen nor Fourier authored the designs). There are similarities, too, in the work and organization of the settlements, and this is where differences appear.

Owen begins from an intention to reform work. This is also the basis of Fourier's (1808) call to reorganize agricultural labour in cooperative units, cited on p. 42. Indeed, the reform of work by the design of better places of work was not uncommon in the early nineteenth century, as at Saltaire in Yorkshire, built for industrialist Titus Salt. Coates writes:

> The whole Saltaire model village was intended to alleviate the conditions of squalor caused by the industrialisation of Bradford. The great textile mill that dominates the village was equipped with a system of flues to provide ventilation. Great plate-glass windows let ample light into the building. Drive shafts for machinery were run between the floors to reduce dust, noise and the risk of accidents and the latest fire safety precautions were installed. Special conduits caught the rain-water and stored it in tanks, ready to be used by the factory. And perhaps most impressive of all the 250-foot mill chimney was fitted with special smoke-burning appliances that meant that little if any of the smoke from the fifty tons of coal burnt each day ever reached the top.
> (Coates, 2001: 187)

Owen and Fourier have moral and practical aims in that suitable environments increase both productivity and well-being. In *A New View of Society* (1813–16; Owen, 1972) Owen proposes a new environment in which a new kind of character could develop, in the belief that family background and material environment conditioned a person's character. He writes:

> for every day will make it more and more evident *that the character of man is . . . always formed for him; that it is . . . created by his predecessors; that they give him . . . his ideas and habits, which are the powers that govern and direct his conduct. Man, therefore, never did, nor is it possible he ever can, form his own character.*
>
> (Owen, 1972: 45, emphasis in the original)

The solution was to provide an environment in which education would eliminate the negative factors in a person's character and replace them by positive ones. He adds, 'all men are irrationally trained at present, and hence the inconsistencies and miseries of the world' (Owen, 1972: 56). It followed they should be retrained, in a modernizing vision that is the principle of nineteenth-century materialism. In Owen's writing it is not the workers, however, who undertake such intervention but the mill owner who has the power, money and educated vision to improve their minds as well as their material and cultural conditions.

Owen wrote the above text more than a decade after being appointed to manage the mills at New Lanark, where he sacked inefficient, corrupt managers and introduced the silent monitor, based on Quaker educationalist Joseph Lancaster's model of negotiated assessment of productivity and character. During an embargo of cotton supplies from North America in 1807 Owen continued – highly unusually for the period – to pay wages. When Quaker buyers were found for the mills, Owen was able to plan for general education so that a new generation of workers might be trained according to his theories. Owen was able to improve workers' housing, and in 1816 opened his Institute for the Formation of Character in a two-storey building providing children's schooling in the daytime and adult education, concerts and dancing in the evening. In his address at its opening, Owen describes its objectives as providing 'immediate comfort and benefit' to all inhabitants of New Lanark, enhancing the welfare of the neighbourhood, with 'extensive ameliorations throughout the British dominions', leading to 'the gradual improvement of every nation in the world' (Owen, 1972: 93). Owen speaks as a prophet, at least as far-reaching in his ideas as Fourier, but is also informed by the ideas of Swiss educator Heinrich Pestalozzi, for whom kindness was more effective than brutality. Owen, I suppose, applies this to his workforce by offering them kinder conditions in which to produce cotton. While, as he says in 1816, he found the workers at New Lanark, 'existing in poverty, crime and misery; and strongly prejudiced . . . against any change that might be proposed' on his arrival there, and while the 'usual principles which have hitherto governed the conduct of men, would have been to punish those who committed the crimes, and to be highly displeased with every one who opposed the alterations that were intended for his benefit', Owen's aim was to produce a new attitude to work, each other and management. New Lanark was to be a beacon of this modern vision, combining design, technology, and education with welfare. Then:

> from this day a change must take place; a new era must commence; the human intellect, through the whole extent of the earth, hitherto enveloped by the grossest ignorance and superstition, must begin to be released from its state of darkness; nor shall nourishment henceforth be given to the seeds of disunion and division among men. For the time is come, when the means may be prepared to train all the nations of the world – men of every colour and climate, of the most diversified habits – in that knowledge which shall impel them not only to love but also to be actively kind to each other.
> (Owen, 1972: 97)

Yet seeded in Owen's ethos of kindness is also his control, as in the self-policing method of the silent monitor. As Vincent Geoghegan remarks, 'Owen displayed both a naïve rationalism and a deeply ingrained elitism', believing that he could persuade the owners of capital to adopt his reformist socialism, based on cooperative principles (Geoghegan, 1984: 100–1). In part, as with most reformism, the attraction to the owners of capital was that improvements in workers' conditions lessened the chance of strikes. But Owen goes beyond this, in planning a radically new society that he attempted to build in North America after the success of New Lanark. In this, Owen is as detailed in his plans as Fourier, at around the same time (through the 1810s to 1830s). But I need to go further.

In specifying buildings for a new community, Owen includes separate schools for infants and older children, public kitchens and communal dining rooms, a prayer room, a lecture room, a committee room, a library, outdoor recreation spaces, and housing for family units of two parents and two children (Coates, 2001: 75). To enforce the ideal family unit, all children over three years old in excess of two per family are accommodated in a dormitory. Centrally located apartments house superintendents, clergymen, teachers and doctors, as if in a panopticion (Foucault, 1979).

Owen's plans to extend from New Lanark to more ambitious settlements in Britain were not realized, though he played a major role in establishing the cooperative movement. In 1824 he was offered the site of Harmony in Pennsylvania, from which the Rappite community (see Case One) had moved to Indiana. The cost of land was lower than in Britain, and the site had most of the buildings needed for, and had shown the success of, a self-sufficient economy. Owen bought the 20,000-acre site for 600–700 people in 1825. Leaving his son William to organize the estate, he went on a fund-raising tour, returning with 'a hand-picked group of naturalists, academics, Pestalozzian teachers and their pupils' (Coates, 2001: 84). But divisions followed and factions negotiated their departure to semi-separate parcels of land.

In Britain, the Owenite movement spread (Coates, 2001: 77–94). A series of National Co-operative Congresses in the 1830s debated the alternative tactics of grand schemes as proposed by Owen and an incremental approach in local projects. One outcome was the community of Manea Fen, Cambridgeshire, on the 200-acre estate of Methodist preacher William Hodson. Coates writes, 'the "Hodsonian Experiment" became something of a rallying point for dissident Owenites critical of the central committee's [of the National Community Friendly Society, founded by Owen in 1837] paternalistic pronouncements. This antagonism increased sharply following the commencement of the official Owenite scheme at Harmony Hall' (Coates, 2001: 92).

At Harmony Hall, built in 1839, Joseph Hansom – architect of Birmingham Town Hall, founder of *The Builder* magazine and inventor of the Hansom cab – designed a three-storey mansion in red brick with library, dining hall, schoolrooms, baths, offices, stores and single bedrooms for apprentices. This Owenite phalanstery, so to speak, was equipped with new steam-powered heating and ventilation systems, even a miniature railway taking dishes to the kitchen. Evening classes were given for the local community in drawing, grammar, geography, agriculture, elocution, music and dancing; but Coates notes, 'the colonizts themselves survived on little more than subsistence rations' (Coates, 2001: 96). Owen resigned from the project in its early stages, returning after it endured a sequence of financial crises to resume the colony's management, 'extending the farms and setting up a fee paying school with fees so high that no working class Owenite could possibly afford them' (ibid.). The project failed again, and was dissolved in 1845. In 1902 the building, after becoming a lunatic asylum and a college, burned down. Which of Fourier or Owen, then, was the more practical utopian? The question is rhetorical for obvious reasons, but the extent of Owen's practical abilities remains in doubt after New Lanark.

Harmony Hall exhibits a typical mix of modernization and Owen's authoritarianism, though the breakaway project at Manea Fen also failed (for financial reasons). Perhaps, had Fourier persuaded Napoleon Bonaparte to be the true Liberator, the phalansteries would have met a similar end in disunity and disorganization, despite the detail in which they had been planned. But to return to the question as to whether Fourier and Owen have similar approaches, there are some significant commonalities. Both Owen and Fourier specify the education of children (though Owen makes them work in the mills as well). Both are obsessive in the detail they provide to add plausibility to their vision. And both look towards a new society, radically different from the old, based on a concept of human well-being. Perhaps here, though, is the essential difference, in that Owen believes in a moral order in which conditions objectively produce right behaviour, designed as a cooperative form of labour overseen by an elite. Fourier retains class difference yet is markedly more egalitarian. His society is based on mutual attraction not on re-education, and his discovery is the analytical tabulation of the passions. Owen does not envisage the erotic life as central to the phalanstery and does not see work as libidinal, though he does encourage dancing. Although Fourier and Owen begin from efforts to see how collaborative labour (industry for Owen, agriculture for Fourier) can be more productive as well as beneficial, the development of Fourier's vision in its libidinal aspect appears unrelated to anything in Owen's reformism and could not have been derived from it. In the end, Fourier and Owen were eclipsed by a diversity of practical utopian settlements through the nineteenth century, in the context of the rise of organized urban and rural labour, the spread of literacy, and the benefits of reform that improved urban and rural conditions of work and housing.

David Pepper notes French nineteenth-century communes 'associated with socialism, anarchism and the political left' (Pepper, 1991: 27). In Britain, too, there were anarchist communes such as Clousden Hill near Newcastle, where in 1895 the Anarchist Land Colony bought 20 acres of farmland for a settlement based on common ownership, mutual assurance, mental and moral improvement, education, productivity in intensive agriculture, and 'the attainment of a greater share of the comforts of life than the working classes now possess' (Coates, 2001: 203–4). This reflects Peter Kropotkin's ideas, set out in *Fields, Factories and Workshops*, first published as a series of magazine articles between 1888 and 1890 (Kropotkin, 1919; see Guérin, 1998: 227–77). But after some expansion, division between factions for free and organized ways of working at Clousden Hill caused a split. Coates relates, 'The attempt to reach consensus on the way ahead resulted in a seemingly endless round of increasingly acrimonious meetings, with the only thing that everyone could agree on being to ask visitors to give a week's notice of their arrival' (Coates, 2001: 205). This might support Owen's managerialism had its outcome not been the same. But my initial question was whether Fourier's texts might best be read as literary works, in the same way as More's. This is clearly not Owen's intention in his writing, and was not Fourier's either – yet there is a literary quality to Fourier's accounts of life in the phalanstery that could be read as a critical foil to the corruptions of bourgeois society against which he reacted. I find, after this sketch of a practical utopianism, that Owen's specification and tendency to modernization, and Fourier's emphasis on pleasure and the libidinal character of a just society, are almost opposite polarities of an axis of reform: neither is revolutionary in the sense of supporting popular insurrection and both, in different ways, seek radical change by constitutional means. But I wonder if that axis needs another point, to become a triangulation in which grassroots movements for the empowerment of a mass public also have a place. To add this, I look now to the Chartist Land Company as a revolutionary rather than reformist (if still constitutional) campaign.

The Charterville allotments

The Chartist Land Company was inaugurated by Feargus O'Connor at a Chartist conference in London in 1845, aiming to buy land on which to establish communities of smallholders. Capital was to be raised by subscription, rents then providing future capital, while plot-holders would enjoy a better and more independent income from land than from wage labour. O'Connor was a persuasive speaker (Ashton, 1999: 65), and his condemnation of land monopoly found a sympathetic public. F. C. Mather writes that while the idea is now overlooked, in its time it was 'a good deal less impracticable . . . than historians habituated to a more advanced industrial economy have been willing to concede' (Mather, 1965: 15). Mather notes that some settlers remained for as much as four years, while the failure of the project was due to both a lack of capital and poor management on O'Connor's part. For Coates, its lack of a legal basis was its undoing, with too rapid expansion.

One of the settlements was at Minster Lovell, Oxfordshire, set on high and stony ground. The plot boundaries have changed little since the 1840s and most of the cottages built for the Land Company remain (Figure 1), though many have been extended. It is still a quiet place away from the rush of city life, though perhaps in the 1840s it was more busy as the first settlers attempted, with little if any previous experience of working on the land, to derive a living from it. Some were artisans from the south of England seeking independence from wage labour, others were from cities in the Midlands and the North, all picked by ballot from members of the Land Company. The estate at Minster Lovell was purchased in 1846, at the same time as two estates in Gloucestershire, Snigs End, and the Lowbands and Applehurst farms. This followed the purchase of 103 acres of neglected farmland at Heronsgate, Hertfordshire, where the project began. Roads were made, then individual paths to each plot; water was drawn from a spring and a well dug. Oak timber was bought for roofing, and deal for walls, while the plots were cleared of roots and stones. Ten canal boatloads of horse manure were brought from London, and supplies of firewood and seeds distributed. Each cottage had a privy, and shared animals and fruit trees. This was a generous provision, contrasting starkly to the conditions of the migrant urban poor described by Friedrich Engels from his research in Manchester in 1844 (a year before the foundation of the Land Company):

> Immediately under the railway bridge there stands a court, the filth and horrors of which surpass all others . . . Passing along a rough bank, among stakes and washing lines, one penetrates into this chaos of small one-storied, one-roomed huts, in most of which there is no artificial floor; kitchen, living room and sleeping room all in one. In such a hole, scarcely five feet long by six broad, I found two beds . . . Everywhere before the doors refuse and offal.
> (Engels, 1892, quoted in Donald, 1999: 35)

In comparison, the Chartist Allotments were paradise – a land of milk and honey.

Subsistence was not the primary aim, however. Just as twentieth-century plot-land dwellers saw their plots and self-built houses as 'a living symbol of freedom and independence' (Hardy and Ward, 2004: 12), so the Chartist Land Company's settlers felt liberated from the effects of rapid urban growth and industrialization. Perhaps more important, though, was that ownership of property was the condition, in England in the 1840s, for the vote (though only for men). As Coates summarizes, 'Land ownership would give working people the vote leading to a new government and then universal suffrage' (Coates, 2001: 101). This represented a departure from the Chartists' initial aim for universal suffrage through campaigning, influenced by contact with groups in North America where adult male suffrage had not ended land monopoly. Jamie Bronstein suggests, 'perhaps economics had to underpin politics and not the other way around. These . . . experiences helped to intensify the Chartists' already evident shift from political to social solutions like the Chartist

Figure 1 Charterville Allotments, Minster Lovell, one of the cottages

Land Company' (Bronstein, 1999: 148). In June 1846 200 people were working at Heronsgate and as Coates writes, 'there was a continual stream of visitors wanting to see for themselves this little piece of paradise taking shape in the Hertfordshire countryside' (Coates, 2001: 103). It failed. A rush of applications for membership of the Land Company caused O'Connor to be lax in keeping records, and the need for capital to buy more land led him to attempt to set up a joint stock company to legalize tenure. But this required the signatures of every stakeholder – an impossible task – and his inability to resolve the situation led to a parliamentary enquiry that exposed his lack of appropriate records. By then some plot-holders were refusing to pay rent, and O'Connor resorted to evictions. Coates continues, 'At Snigs End settlers told the bailiffs . . . they would "manure the land with their blood before it should be taken from them" . . . some managed to claim legal title to their holdings, others ended up at the mercy of the Poor Law Guardians' (Coates, 2001: 105). The Land Company was dissolved in 1849. In 1852 O'Connor was taken to an asylum, and died three years later. Coates argues that had the experiment been successful, 'a whole new pattern of land ownership would have been established and perhaps had working people gained the vote . . . a completely different history of the labour movement . . . would have followed' (Coates, 2001: 105). Dennis Hardy and Colin Ward see the allure of the land, generally, as restoring common rights in the tradition of medieval peasants' revolts and the Diggers (Hardy and Ward, 2004: 11). I would add only that the basis of the scheme appears viable, if compromised by a lack of suitable land-working skills and the absence of a legal structure. But the Diggers did not have the latter either. Power is not donated, only taken or dissolved.

Utopia as fiction or programme?

The Chartist Land Company may be a precedent for today's campaigns for social and environmental justice, as the Diggers were a precedent in name at least for radical groups in the 1960s (see Chapter Five). The Land Company failed, but not because the core idea was implausible. In the triangulation, then, that I proposed above, the axis of utopian visions is redrawn in a tripartite arrangement, the third point of which is practical utopianism such as repossessing the land, or today direct action. Beside that, Owen's utopianism seems managerial, in the British context not unlike the New Labour modernization that is one offshoot of the movement of which Owen's cooperative principle was a foundational (if now overlooked) element. Fourier, I think in contrast, seems, the more I reconsider his texts, closer to More's literary utopianism, his description of the phalanstery and its daily life a foil to the miseries of real life in a bourgeois state, and having begun in his first texts as satire.

The life of the phalanx involves several kinds of work within the day, extensive and diverse socializing, amorous expression and occupancy of buildings like palaces. Marx found Fourier's idea of attraction frivolous, developing instead a theory of alienation and class struggle. Both Marx and Fourier, however, saw their work as scientific. Fourier's texts seem prescriptive, particularly if subsumed in Marxist scientific socialism, but I would say they are not if read as metaphorical or critically distanced. There are still issues, not least in the role of Fourier as priest-like forerunner of a new society to which his writing will lead the mass public, though via the agency of an enlightened ruler. The erotic life of the phalanstery is set out in detail but the power relations by which it is to be brought into being remain either vague or reactionary. Michel Foucault cites a letter of 10 August 1836 in the Fourierist paper *La Phalange*:

> Moralists, philosophers, legislators, flatterers of civilisation, this is the plan of your Paris, neatly ordered and arranged, here is the improved plan in which all things are gathered together. At the centre, and within a first enclosure, hospitals for all diseases, almshouses for all types of poverty, madhouses, prisons, convict-prisons for men, women and children. Around the first enclosure, barracks, court-rooms, police stations, houses for prison warders, scaffolds, houses for the executioner and his assistants. At the four corners, the Chamber of Deputies, the Chamber of peers, the Institute and the Royal Palace. Outside, there are the various services that supply the central enclosure, commerce, with its swindlers and its bankruptcies; industry and its furious struggles; the press, with its sophisms; the gambling dens; prostitution, the people dying of hunger or wallowing in debauchery, always ready to lend an ear to the voice of the Genius of Revolutions; the heartless rich . . . Lastly the ruthless war of all against all.
>
> (Foucault, 1979: 307)

This seems an ironic representation of a city framed in the vocabulary of the phalanstery. For Foucault it denotes a coercive surveillance, a city organized by rules and institutions to replace and replicate by other means the previous culture of public bodily punishment; I read it more as a satirical explanation based on an assumption, which is not satirical, that the built environment denotes and produces a society's ethos.

Fourier gives little attention to agency. As Barbara Goodwin writes, 'Ultimately we are left with a puzzle: Civilization is clearly an unnatural yet self-sustaining system, but Fourier gives no satisfactory explanation of the *origins* of this deplorable state' (Goodwin, 1978: 15). Fourier explains that civilization is based on repression but omits to say how people will free themselves from it. For Francis, 'Having discovered the blueprint for utopia, . . . [Fourier] was famously uninterested in trying to put it into practice, believing that a partial attempt would challenge the integrity of the structure, expecting that when the time was ripe, the phalansterian series would somehow come into existence' (Francis, 1997: 77). Other French nineteenth-century utopians sought to transform society through the agency of specific groups. For Henri de Saint-Simon it was through art and science, as he wrote in 1819:

> The prosperity of France can only be achieved through the progress of the sciences, fine arts, and arts and crafts. Now, the princes, the great officers of the Crown . . . and the idle property owners do not work directly for the progress of the sciences, fine arts, and arts and crafts . . . they endeavour to prolong the supremacy hitherto exercised by conjectural theories over positive knowledge.
> (Carey, 1999: 186–7)

Saint-Simon gives direction of the state budget for agriculture to farmers, for commerce to merchants, and so forth. A Chamber of Invention of 300 engineers, poets, painters, architects and musicians oversees public works and festivals. A Chamber of Examination comprised of physicists and mathematicians scrutinizes its work and makes plans for public education, religion and more festivals. A Chamber of Execution consisting of industrialists supervises the carrying out of projects and collection of taxes (Carey, 1999: 184–92). This echoes Fourier's accommodation of class difference, but Saint-Simon privileges an artistic avant-garde (Nochlin, 1968: 5). Pierre-Joseph Proudhon, too, wrote in 1865 that art could expose society as it is: 'In order to improve us it must first of all know us, and in order to know us, it must see us as we are' (Edwards, 1970: 215). Proudhon rejects Fourier's systematic approach (Buber, 1996: 25), regarding the authority of one person over another as in inverse ratio to the intellectual development of the society of which they are members, so that in an advanced society the extent of power is minimal.

Then, 'The probable duration of that authority may be calculated according to the more or less widespread desire for true government, that is, government based on science' (Edwards, 1970: 88). The force of will gives way to that of reason, while 'Property and royalty have been decaying since the world began. Just as man seeks justice in equality, society seeks order in anarchy' (Edwards, 1970: 89). A fundamental question in all this, that affects the appropriate reading of nineteenth-century utopian texts as either plan or literary foil to present reality, is whether the means employed are pragmatic (or borrowed from the dominant society), are means towards an end, or are the end itself, enacting it in embodied values and appropriate practices.

Alexander Herzen (1979; 1980) is cited by Colin Ward: 'What is the purpose of a child?' he asks. 'We think that the purpose of a child is to grow up . . . But its purpose is to play . . . to be a child. If we merely look at the end of the process, the purpose of all life is death' (Ward, 1991: 50–1). Martin Buber does not mention Herzen in *Paths to Utopia* (1996), and Adam Ulam, in an essay on socialism and Utopia, mentions only his sympathy for the Confederacy in the Civil War (Ulam, 1973: 120). But Ward cites this passage from Herzen's *From the Other Shore*: 'An end that is infinitely remote is not an end, but a trap; and end must be near – it ought to be . . . the labourer's wage, or pleasure in the work done. Each age, each generation, each life had and has its own fullness' (Ward, 1991: 60). From this it follows that the means *are* the ends.

Marius de Geus notes a criticism of utopian socialism, which is that 'utopian thinkers generally lack a thorough analysis of the power relations within society, the economic developments which prepare the way for changes, and the groups who desire changes' (de Geus, 1999: 47). Goodwin writes, 'Fourier's society is organized so that disruptive tendencies do not arise: the liberated passions are never anti-social, and all potential conflict is defused by multiple allegiances' (Goodwin, 1978: 117). If I read Fourier as writing in a textual gap between plan and fiction, between prescription of a new society and a dream of another world moulded by critique of his own society, then his work draws attention to the oppressive basis of civilization later discussed by Freud (1946), and Marcuse (1956). I find the idea of a libidinal society interesting in a way that now informs my thinking about activism in the late twentieth century (see Chapter Five), and perhaps opens a possibility other than the contestation of power that leads too often to an impasse, or the replication of power in the reformism of elites. If utopianism is the pursuit of dreams, the gap between dream and reality is not only the temporal space of a trajectory from yesterday to a tomorrow that tends to reproduce it in dreamy form; it is also a space in which to recognize the (erotic) creativity of ordinary life.

Part Two
Practices

4 New cities

Introduction

In the first part of this book I reconsidered literary, historical and theoretical Utopias from the sixteenth to nineteenth centuries. In Chapter Three I charted a shift from the literary and metaphorical intentions of early utopian texts to a practical ambition to build a new society, but argued that Fourier's phalanstery is more illuminating if read as a literary model. Continuing objections to utopian thought are that it relates only to marginal or elite groups, and that it takes geographically marginal forms in remote rural settlements. In this chapter, then, I consider utopianism on an urban scale, and investigate the role of reformism, philanthropy and professional expertise in relation to a need for new models of power. I move also into the technological period of the mid-nineteenth to the early twenty-first century.

The chapter is structured around four cases, though I investigate these discursively rather than descriptively. I begin with Idelfons Cerdà's (1859) plan for the northern extension of Barcelona, then move to the Garden City proposed by Ebenezer Howard and Le Corbusier's plans for a re-engineering of Paris and Algiers. Finally, to bring the chapter into the present, I comment briefly on BedZED, a suburban housing project aiming to pilot sustainable urban living in the face of insurmountable evidence that climate change is produced by human action.

The northern extension of Barcelona

Barcelona is now a major destination for cultural and business tourism, with world-class museums, a wealth of architectural monuments from the *modernista* (art nouveau) period, contemporary art museums and new public spaces created in preparation for the city's hosting of the 1992 Olympic Games (Degen, 2004; Miles, 2004b). From the mid-nineteenth century to the late 1990s its planning regime consisted primarily of regulation for public benefit. The city's northern extension – Eixample – is a quiet, mainly middle-class district of tree-lined avenues, small restaurants and fashionable shopping streets. This is the city's extension based on Cerdà's 1859 plan and subsequent revisions. The infrastructure and street grid, the proportions of streets and urban blocks, and provision of garden squares within each block remain largely as specified by Cerdà, though he left the city before the building of Eixample was complete, following a series of altercations with the city's

60 *Practices*

government. But while factories were built in peripheral sites rather than within the extension as Cerdà proposed, and there is less green space, Eixample remains a case of humane urbanization – a plan to build a new city district on the basis of public benefit in order to create the conditions for a more orderly society.

Cerdà believed in the necessity of planning as a means to ensure civil order, and in the rational plan as stating the liberty promised in the European Enlightenment. He saw all social classes as deserving well-ordered spaces in which to live orderly and satisfying lives, and writes in *Pensiamento económico* (1860), 'the *square* grid is justice itself and equality of rights, to which every professional who may design projects for building, reforming or extending a town should aspire' (Soria y Puig, 1999: 127). This has two contexts. First, on 11 September 1714, Barcelona had fallen to the Bourbons. The city was classified as a military zone and no building was allowed in proximity to its ramparts. The lack of outward expansion meant a worsening of housing conditions, and limited the prospect for commercial growth. The city's merchant class called for the demolition of the walls in 1840, and established a commission to map an extension in 1846 (Gimeno, 2001). The demolition of the walls was sanctioned by the Madrid government in 1860, in part through Cerdà's negotiations, as a result of which he was seen in Barcelona as close to the loathed Madrid regime. In response, the city authorities organized a competition for plans for the city's extension and chose a rival proposal. Then Cerdà's plan was imposed by Madrid, as a result of which he remains suspect among the planners in Barcelona even today. To me his plan was the most radical, even though most competition entries employed either a grid or radial plan. Cerdà's plan (after a preliminary study of 1855), however, adopted a grid of equal proportions for the entire zone, and equated it with an ethical quality.

This brings me to the second context. Conditions in the old city produced repeated outbreaks of contagious diseases combined with unrest, as in a major strike in 1855. Cerdà's *Statistical Monograph on the Working Classes of Barcelona in 1856* was commissioned in response to both eventualities. Cerdà studied housing, wage levels, life expectancy and dietary standards, taking 333 occupational categories. He found that adequate dietary provision occurred in only 11 categories, and that life expectancy among workers was half that of the bourgeoisie. If, then, order and growth required a more equitable treatment of citizens, an extension provided the opportunity to deliver a new urban environment. Cerdà proposed decent housing for all classes in villas and apartment blocks, access to green space at a minimum distance for all, clean air, street lighting and seating, provision for hygienic markets and rapid public transport. Workers were to live near factories, artisans above shops – resembling in more progressive terms the model of an urban village that figures in urban renewal strategies today. The plan was aesthetic and ethical, and these aspects are inseparable. But it is not revolution, rather liberal reform aimed at curbing unrest and promiscuity as well as disease, by removing the conditions in which, he recognized, they were produced.

Cerdà's underlying concern was for civil order in an alignment of technology and urban spatial practices for a new social organization. Arturo Soria y Puig summarizes:

The present . . . was the result of a historic process, moulded by diverse forces. By analyzing their vectors he would be able to assess the breaks with the past and use new tools. He pinpointed the characteristics of the new era in three areas . . . It was not a matter of political, religious, moral or behavioural change; these transformations were to come later. It was a total break, powered by the new source of energy, steam. . . . The first consequence of the availability of more power . . . was mobility . . . [His] second insight [was] the railway . . . [which was] about to turn the city upside down . . . His third piece of farsightedness was even more amazing . . . The telegraph had been invented, but . . . its use was still extremely limited. His deduction . . . that communicativeness was one of the essential features of the new civilization must be seen as a milestone.

(Soria y Puig, 1999: 15–16)

The three areas – steam power in place of animal power, rail transport in place of the cart, and telecommunications – are integral to a city of viality, as Cerdà terms what might today be called liveability. Soria y Puig notes that Cerdà's training was in the Cartesian tradition of an engineer which led him to 'a sometimes dramatic pursuit of order, harmony, and even symmetry' (Soria y Puig, 1999: 17). It led him also to accept that contradictory needs can be combined, so that the non-predicable behaviour of a large population is complemented by the ordering principle of the regular grid. Similarly, 'Any qualified professional entrusted with . . . the reform, extension, or founding of a city, must consider the street both as a courtyard or forecourt to the house and as a public way' (Soria y Puig, 1999: 111). This enables families to use the spaces in front of dwellings as extensions of domestic space while the movement of traffic is unimpeded. In the *General Theory of Urbanization* (*Teoría general de la urbanización y aplicación de sus principios y doctrinas a la reforma y ensanche de Barcelona*, 1867), Cerdà argues that because carriages cause congestion when they stop, margins and loading bays for goods are needed, but should be located in transverse streets while 'viality in the longitudinal direction needs . . . lack of encumbrance' (Soria y Puig, 1999: 191–4). He specifies the proportions of buildings and widths of streets – half the street for pedestrian use – together with the details of facades, and a special type of urban block for sites adjacent to the railway, with basement deliveries, ground floor shops and apartments above (Tarragó, 2001). All this depends on regulation, though Cerdà separates the role of public administration from that of management and delivery, the latter effectively carried out by the private sector.

In the *Theory of Urban Viability* (*Teoría de la viabilidad urbana y reforma de la de Madrid*, 1861), Cerdà writes:

Successful planning requires . . . the circumstances of the locality, its centres of life, the needs of general viality, and the number, habits, and customs of the population should be studied. Justice demands that the interests of each and every one be attended to and respected with all possible equality. General convenience insists that social and political life, mercantile, economic, and industrial life, be allowed the resources they need . . . to function.

(Soria y Puig, 1999: 125)

It is not, I think, a question of one set of needs following another, but of an integration of the needs of all sections of a modern urban society in ways conducive to commercial growth and which ensure social justice. Cerdà was horrified by the conditions he saw in researching his 1856 report, and worried at the prospect of insurrection after revolts in other European cities in 1848. Looking at precedents in Latin America he saw a general absence of such unrest and of armed incursion despite a lack of fortifications, writing in 1859 that the barricades built in narrow alleys in Paris were unknown in Buenos Aires (Soria y Puig, 1999: 135). There may have been a number of reasons for this. The point, though, is that Cerdà combines a desire for public well-being extended to all classes with a requirement of orderly behaviour on the part of citizens, assuming that it is produced by orderly spaces under a benevolent planning authority. Cerdà's 1859 plan remains a progressive document, contrasting with the remodelling of Paris by Baron Haussmann in the 1850s and 1860s. It contrasts also with Howard's Garden City in that Cerdà at no point gives up on the urban as the site of a potentially better world, celebrating steam power, rail transport and telecommunications as beneficial means to urbanization.

Garden City

Cerdà's idea of urbanization revolves around built order as conducive to social order, applicable to new towns and urban redevelopment alike. Today the planning authority in Barcelona describes its projects for the improvement of public spaces *within* the city as urbanization. In England, in contrast, urbanization tends to mean extension of the city in the green belt, reflecting a town-and-country duality that is anti-urban and has its roots in a negative reaction to industrialization on the part of late nineteenth-century elites, for whom the countryside is a residual Eden. The city breeds crime, disease and disorder, in a characterization that conveniently overlooks the real conditions in which the urban poor dwell (Engels, 1892). In nineteenth-century London the threat of disorder was taken seriously but addressed not, for the most part, through social reorganization but through cultural measures, including the creation of unifying national myths and the establishment of art and ethnographic museums. At the same time, for the Romantic movement, the city is a site of anonymity, while for early sociologists it is a location of new contractual forms and synergies possible only in the conditions of proximity (Miles, 2007: 8–21).

By the 1880s the worst aspects of industrialization were mediated and improvements such as sewers had ensured better public health. Reform produced advances in working-class education, new kinds of housing in large blocks, such as those around the site of the new art gallery funded by Henry Tate on the south bank of the Thames, on the site of a former prison (Taylor, 1994). But most importantly for this chapter, the metropolitan railway system and growth of cooperative building societies allowed clerks, shopworkers and servants to leave the inner city to live in suburbs. Richard Sennett writes:

> the working poor who could scrape together the money could now move away from the city center to individual homes of their own. As in housing for the privileged, these modest row houses consisted of uniform blocks, with individual small yards and outhouses in the back. ... By working-class standards ... the housing was an immense achievement. People slept on a different floor than they ate; the smell of urine and faeces no longer pervaded the interior.
>
> (Sennett, 1995: 334)

Sennett refers to the progressive erasure of odours, the growth of mass-produced housing and development of a suburban lifestyle which meant that, 'By the 1880s, the urban tide which had flooded into London began to flow out' (Sennett, 1995: 334). This is the context for the Garden City, first proposed by Howard in the 1890s when the middle classes, too, were moving to suburbs such as Golders Green where houses had large gardens, almost like living in the fabled countryside, and where they could commute to work or visit the city's department stores by underground railway. Howard's plan extends this outflow to further locations, but within the grain of an already evident de-urbanization.

Howard derived the idea of a Garden City from North American and British colonial models, and had seen the rebuilding of the centre of Chicago after the 1871 fire. He knew of Edward Gibbon Wakefield's plan for South Australia, James Silk Buckingham's for a model town for 10,000 people (Hall and Ward, 1998: 12), and the philanthropic industrial villages of Bourneville (1879) and Port Sunlight (1888). Howard synthesized aspects of all these to propose a new, small city in the countryside. In *Tomorrow: A Peaceful Path to Real Reform* (1898) – republished in 1902 as *Garden Cities of Tomorrow* – he argued that cities suffer from overcrowding, pollution, absence of nature, the expense of sewers, and alcoholism, but attract migrants seeking well-paid work and amusements. Rural life is healthy yet offers little employment and that at low wages. His solution is to bring the best aspects of the two together, which is naïve but was credible in the 1890s when rural land values fell, making the cost of a suitable estate relatively affordable at a time when, for the same reason, rural landowners did not invest in new housing for agricultural workers.

Howard is as detailed in his specification of Garden City as previous utopians. It was to have a population of 30,000, with a further 2,000 in a surrounding agricultural belt, and 5,500 houses on a 5,000-acre site, the average size of a plot being 20 by 130 feet. At nearly six people to a house this implies what would now be large families but in 1898 this was not exceptional. More to the point is that the outer ring, marked by the Inter-Municipal Railway, was only three-quarters of a mile from the centre, which consisted of a five-acre garden surrounded by public buildings including a town hall, museum, art gallery and library. Garden City was to be a city with a garden in the centre, surrounded by a green belt, and the walk from the outer edge to the centre would take 15 minutes. Various institutions were to be relocated to Garden City from London, such as reformatories and convalescent homes that might benefit from clean air. Schools and playing fields would be sited along a ring road called Grand Avenue. Industries such as jam making, clothing, and light engineering would provide jobs, use locally produced raw materials and meet local needs. Houses would be clustered in six sections divided by radial boulevards on the far side of Grand Avenue, within sight (across the railway) of pastures and fruit farms, but also brick factories and an agricultural college (illustrated in Hall and Ward, 1998: fig. 6). Peter Hall and Colin Ward note, 'Howard believed that industrialists would gladly follow the lead already set by pioneers like Cadbury at Bourneville and Lever at Port Sunlight; they would see the advantages of operating in a clean smoke-free atmosphere' (Hall and Ward, 1998: 20). Hall and Ward see the Mall in Washington as a model for the central park, but regard Howard as original in proposing a glass-roofed winter garden on its perimeter, citing the following from an unpublished article by Howard:

> Running all around Central Park is a wide glass Arcade or Crystal Palace. This building is in wet weather one of the favourite resorts of the people, for the knowledge that its bright shelter is close at hand will tempt people into the park even in the most doubtful of weather. Here manufactured goods are exposed for sale, and here most of the shopping which requires the job of deliberation and selection is done. The space is, however, a good deal larger than is required for these purposes, and a considerable part of it is used as a winter garden, and the whole forms a permanent exhibition of a most attractive character – the furthest inhabitant being within 600 yards.
> (Hall and Ward, 1998: 21, citing Beevers, 1988: 52)

There were by then glass-roofed arcades in London and Leeds as well as Paris, used for goods requiring deliberation – luxury goods rather than vegetables. Crystal Palace in London is another precedent, and winter gardens were popular in seaside resorts by the 1890s. But Howard unites the two, bringing the leisure of shopping and the leisure of strolling among plants under one circular, glass roof. I recall walking through the mall at Milton Keynes, seeing an oak tree at the centre, and wondering how much of Howard's progressive urbanism had survived in what, arguably, is one of his legacies (see Figure 2). But, though Hall and Ward see Howard as foreseeing the mall, I read the glass arcade more as a site of social mixing like the Mall in Washington DC (Sennett, 1995: 269). Despite its rural site, Garden City had several explicitly modern, urban qualities derived from North American models (including the numbering of avenues).

In *Tomorrow: A Peaceful Path to Real Reform* (1898), Howard sketches an interesting idea: that falling land values enabled purchase of the land with moderate capital while its development would increase land values. If initial capital could be raised from philanthropic sources, then the increase in land values combined with accumulated rents as people moved to Garden City could repay the capital and provide a welfare fund which would enable Garden City to become self-

Figure 2 Milton Keynes, the oak tree in the mall

administering without government involvement, projected as happening 30 years after the initial purchase of land. To implement this, Howard set up the Garden City Association in 1899, which became First Garden City Limited in 1903. He hoped the Cooperative Society would build Garden City – or Social City, as a larger plan with satellite cities linked by an inter-municipal canal system to house a quarter of a million was called in the 1902 edition of his book – but the purchase of a site required capital which was forthcoming only from industrialists. When a Garden City was begun at Letchworth, only £40,000 of a required £300,000 had been raised, all from the company's directors (Hall and Ward, 1998: 34). The idea that Garden City would be self-managing was replaced by an orthodox financial model including a duty to shareholders; and while Howard proposed a mix of architectural styles, the architects of Letchworth, Raymond Unwin and Barry Parker, introduced their enthusiasm for medievalism. Although a pilot scheme in 1905 demonstrated the viability of low-cost agricultural housing in modern materials, the new town was designed for middle-class families. Rents were too high for most rural workers, who cycled in to jobs from surrounding areas. Elizabeth Wilson remarks that Letchworth's inhabitants were middle-class radicals – 'vegetarians, dress reformers and eccentrics' (Wilson, 1991: 102). She cites Ethel Henderson, who moved there in 1905: 'we were looked on as cranks . . . I suppose we were cranks . . . Bare legs and sandals for both men and women' (Wilson, 1991: 103). Wilson concludes that Letchworth is a 'sanitised utopia, with the obsessional, controlling perfection that characterised all utopias' (Wilson, 1991: 104). Chris Coates takes a more positive view, seeing Garden City as part of a history of the cooperative movement, inspired by Edward Bellamy's *Looking Backwards* (1888; see Carey, 1999: 284–93; Coates, 2001: 209).

Howard, though, admired Russian anarchist Peter Kropotkin (Hall and Ward, 1998: 28) whose *Mutual Aid* was published in England in 1902 and whose writing appeared in magazines during the 1890s. The autonomy of Garden City reflects Kropotkin's idea of cooperation in small communities, and its welfare ethos also reflects North American land reformer Henry George's *Progress and Poverty* (Meller, 2001: 125). Perhaps the idea of autonomy is rooted in North American precedents, in the development of the new social sciences of sociology and economics and in the absence of mention of cities in the Constitution. A task of social scientists was to establish an identity for the city by, as Thomas Bender writes, calling for the autonomy of city governance on quasi-scientific grounds. He adds, citing Mary Ryan (1997: 128), that 'an inclusive and robust democracy' existed in cities before the 'crisis of union' while a new confidence in urban governance emerged in projects such as New York's Central Park (Bender, 1999: 27; see also Gandy, 2002). These ideas provoked attacks on universal suffrage and the extent of civil powers, while vital services were still managed by state bureaucracies (Bender, 1999: 28), but 'Home rule became a rallying cry that united urban populations; the need for political authority was great enough to produce a cross-class collaboration. . . . The . . . needs of urban populations prompted the invention of a social politics' (Bender, 1999: 29).

This puts Garden City in another context – that of city administration as distinct from state government as agent of improvement. The difficulty, echoing that of the Chartist Land Company (Chapter Three), was that Howard never thought through the implementation of the idea of a self-managing city. Hall and Ward see Garden City's Central Council (as landlord for the community), Departments (of policy and public works) and limited company (to distribute dividends to shareholders) as always likely to come into conflict. Perhaps, had Garden City been managed as proposed by Howard, they would have come into conflict, too, with central government.

The title of Howard's book, *Tomorrow: A Peaceful Path to Real Reform*, published a year after Charles Booth's 1897 report on poverty in east London, is significant. The urban lower and middle classes were already moving to the suburbs, and Garden City was meant to extend the drift to a further – though still urban and potentially metropolitan – environment. This rural-urban idyll has a moral more than a material quality, and Howard, like Cerdà, was concerned for social order, extending the moral aspect of Bourneville – where the behaviour of workers at the Cadbury factory was regulated, for instance, in separate recreation spaces for men and women – to the construction of an environment of good citizenship. A small-scale settlement had advantages in this over a large city. Hugh Barton writes that after Robert Owen's New Lanark it was assumed 'that the village, or something like it, was the ideal settlement for human beings to inhabit, and hence one should seek to create to replace large cities' (Barton, 2000: 21). He cites Tony Scrase (1983) on the way that nineteenth-century Utopias are romanticized, 'showing a preference for small, self-contained communities and allowing the inhabitants to work for at least part of their time in agriculture or their garden' (Barton, 2000: 21). But this could mean a number of things and need not rule out other advantages of small settlements in terms of self-organization and social ownership. First, it could mean self-organized living in self-sufficient communities, as in the ecovillages and intentional communities discussed in Chapters Five and Six and Cases Seven to Nine. Second, it could mean areas of self-organization in a larger urban environment, as in self-build schemes like those pioneered by Walter Segal (Broome and Richardson, 1995: 73–86; Vale and Vale, 1991: 135) or home production of food on urban allotments (Crouch, 2003). Third, it might mean freedom from large structures and systems of governance, as sought by the owners of plot-land cottages in the inter-war period (Hardy and Ward, 2004). If Howard, as Wilson says, produces a sanitized Utopia, and Garden City is de-politicized as well as masked in an aestheticism translated by Parker and Unwin at Letchworth as fake medievalism, this is not the only possibility. A quite different possibility is articulated, for instance, by Canon Barnett in *The Ideal City*, based on his experiences of living in Bristol in the 1890s. I want to draw very briefly on Barnett's text before contrasting Howard's vision with Le Corbusier's.

Barnett writes that an ideal city is a great cosmopolitan seaport where 'there will be no great wealth, because education will have made wealth seem vulgar', and those who have wealth 'will spend it on the city for the citizen's good' (Meller, 1979: 59). These benefactors will live in voluntary simplicity and provide open spaces and the decoration of public buildings, employing artists to 'make the streets a very gallery of pictures' (ibid.). If this sounds like the philanthropy on which Bourneville and Garden City relied, it is mediated by a concern for social justice rather than productivity as the responsibility of the privileged: 'great poverty will also be absent, because great poverty is a sign of neglect' (ibid.). Barnett continues that the poor are poor because others – not they – have erred, schools are inefficient, alcohol is freely available and the conditions in which they dwell breed sickness. There is a moral imperative, in the context of late nineteenth-century reformism and the temperance movement, but the emphasis is on social equity as the precondition for reform, not its eventual outcome. This seems important and a possible criterion for evaluating the ideas and principles of International Modernism, which arose in the inter-war period, two to three decades after Barnett wrote *The Ideal City*, and became a dominant approach based on social engineering in the mass housing projects of the 1950s and 1960s. Within International Modernism there were plural and at times divergent voices, yet the legacy by which it is most often identified today is that of the tower block and the peripheral estate. It may be worth remembering, then, that Barnett's city is peopled, energetic, supportive of a simple rather than an austere life, and open to change. It is also seen from street level.

Le Corbusier

In *Aircraft*, Le Corbusier includes a projected aerial view of his 1933 plan for Algiers below the text, 'Peace among men will come with the undertaking of great works of machine civilization . . . The joy of common action is due rank and discipline. . . . Sweep away the refuse with which life is soiled . . . Let us undertake the great tasks of the new machine civilization' (Le Corbusier, 1987b: plate 110). It sounds like a manifesto. In the case of Algiers it means sweeping away the Arab city to make way for white, curving urban blocks. On the next page is an aerial view of Rio de Janeiro. Le Corbusier writes of the aerial view that it enables cities to 'arise out of their ashes' (plate 96); and of 'a fatality of cosmic elements and events' (plate 122, an aerial drawing of a meandering watercourse in the desert of Boghari). The aerial view is a view from a position of power that provides the overview necessary for grand planning exercises but also introduces the sublime – the feeling of being dwarfed by a vastly powerful other. Le Corbusier appeals repeatedly to a higher order, or unifying force, and if his use of aerial views of landscapes and coastlines (plates 1, 49, 58, 59, 66, 106, 114, 115, 118, 120, and 121 of *Aircraft*), or cities (plates 101–4) contrasts with the machine quality of close-up photographs of aircraft, their shining surfaces may yet constitute another sublime as universalizing in a machine aesthetic as that of the aerial view Sublimity informs the sweeping white blocks of his plan for Algiers and the towers he proposes in *Towards a New Architecture*.

Saying that cities are too dense for security and not dense enough for business needs, he asserts: 'If we take as our basis the vital constructional event which the American sky-scraper has proved to be, it will be sufficient to bring together . . . the great density of our modern populations and to build . . . enormous constructions of 60 stories high' (Le Corbusier, 1946: 57). In *The City of Tomorrow*, he illustrates a cluster of towers surrounded by low-rise industrial units amid fields and waterways under the text, 'A model city for commerce! Is it the fancy of some neurotic passion for speed? But, surely speed lies on this side of mere dreams; it is a brutal necessity' (Le Corbusier, 1987a: 190–1). This offers a number of interpretations. First, the situation of the future city in a green landscape posits a city–countryside relation of total differentiation: city as figure and countryside as the ground on which to inscribe it. Second, the totality of the city is a futuristic projection: an illustration in the magazine *Amazing Stories* (April 1942, in Taylor, 1990: 165) depicts towers in a similar style to illustrate a technologically driven future world. But Le Corbusier's version of a machine age differs in an important respect from Cerdà's celebration of technology, in that the former lacks regard for the specifics of human lives as society is propelled by a sublime Progress. Third, the city is a problem to be solved, as in Haussmann's surgical treatment of Paris as if the fabric offers no resistance. The city is like a machine to be engineered (Donald, 1999: 57); as in a machine, the parts have discrete functions translated as separated work, living, leisure and circulation zones. James Donald writes: 'Behind all this . . . lay an ambition far more audacious than the desire to manage the urban population . . . the construction of a new framework of experience that would determine any possible social behaviour, and so create a new type of person' (Donald, 1999: 59). Renata Salecl, writing on Nicolai Ceausescu's New Bucharest, observes that Le Corbusier's project was a fulfilment of perceived superior demands in a machine age and, in psychological terms, an image of 'the big Other for whom his fantasy was staged; ... Le Corbusier had posited this ... in the place of his Ego Ideal from where he then observed himself in the way he wanted to be seen' (Salecl, 1999: 107–8). I do not have scope to elaborate but I think this version of the sublime sheds light on the element of fantasy in some utopian projects: the total transformation of a world and the absolute difference of the new correspond to the perceived decrepitude of the extant, and this becomes a non-ideal which in its non-correspondence to the ideal cannot be allowed – a model of splitting in the psyche, described by psychoanalyst Naomi Klein (1988: 262–89, 306–43). Claiming a dispassionate view, Le Corbusier proclaims, 'the centres of our great cities must be pulled down and rebuilt' (Le Corbusier, 1987a: 96).

The dispassionate view is a view from a higher moral plane, to be implemented not through philanthropy but through overt power, and Le Corbusier's Voisin Plan for Paris (named after a sponsor) was published in *Nouveau siècle*, the magazine of the French fascist party, in 1927. George Valois, founder of *Le Faisceau*, called it 'an expression of our profoundest thoughts' (Antliff, 1997: 137). Le Corbusier was politically naïve, but an opportunist who saw the power of a centralized state as a viable agent for a plan he saw as more important than his architectural designs.

The other possibility was a colonial regime, and, with his son Pierre Jeanneret, he produced the Obus plan for Algiers for the centenary of French occupation in 1933. A central boulevard sweeps from the hills behind the city to the waterfront on a continuous block with apartments for 180,000 people, cutting through the 'supple-hipped and full-breasted' body of the landscape (Çelik, 2000: 326; Miles, 2004a: 65, n.30). Another curving block sweeps along the waterfront. The Arab quarter, seen as overcrowded, is partly preserved with lower density but only for cultural purposes. Most of the Arab city is erased for open spaces separating white and non-white zones. In *Aircraft*, Le Corbusier says, 'The white race goes its conquering way' (Le Corbusier, 1987b: plate 107), and, as Beatriz Colomina shows, Le Corbusier's conquering way included claiming that a villa designed by Eileen Gray was his (Colomina, 1996: 82–8), in accord with Jennifer Robinson's observation that he attempted 'to claim one of the first international-style buildings in the world as his own work, rather than that of the local Brazilian architects' with whom he worked (Robinson, 2006: 76).

For Le Corbusier, still, the urban block is home to families, albeit white, middle-class French families, for whom he envisages a cell. Fusing the biological and architectural meanings of cell he asks what the needs of human comfort are, and what is necessary to fit together a number of cells in a 'manageable colony in the way an hotel or village is manageable' (Le Corbusier, 1987a: 215, cited in Overy, 2002: 119). He lists liberty, amenities, beauty, economy, low cost, health and harmony; and proposes a site of 400 by 200 metres for a block in which every flat is like a two-storey villa with garden space, facing away from the road. The block overlooks a park with a football pitch, tennis courts, clubhouse, woods and lawns. Cleaning is done by professionals, so 'the servant problem will be solved for you' (Le Corbusier, 1987a: 220). The block is geometrically designed using a principle of repetition to suit mass-production: 'We are unable to produce industrially at normal prices without it' (ibid.). If this resembles systems-built tower blocks, it is in the context of efforts to produce decent urban mass housing, and the work of the International Congress of Modern Architecture (CIAM), founded in 1928, of which Le Corbusier was a member (Curtis, 1999). Whatever else can be said of Le Corbusier, then, the plan of modern architecture has a utopian strand in this aspiration to build new cities that incorporate principles of economy, health, and access to amenities on a mass scale. That this produces the peripheral estates which are not so different whether in North America, western Europe or the Eastern bloc is not inconsequential and requires explanation, but this is not the only aspect of Modernism. The focus of explanation, indeed, might be on why the dream of a city of mass well-being, if articulated by Le Corbusier in terms of the white, bourgeois class, was not delivered.

Some members of CIAM sought radical, participatory solutions. Older members, including Le Corbusier, saw a new society as necessarily engineered by design and professional expertise. As Manfredo Tafuri writes:

> The architect is an organizer, not a designer of objects . . . the search for an authority capable of mediating the planning of building production and urbanism with programmes of civil reorganization is pursued on the political level with the institution of the CIAM. The maximum articulation of form is the means of rendering the public an active and participant consumer of the architectural product.
>
> (Tafuri, 1976: 125–6)

Perhaps the flaw is both in a power-relation of a professional elite to a disempowered population (whom Le Corbusier, in Tafuri's account, seeks to energize and provoke into participation on the terms of striking formal intervention) and in the functionalism of a machine aesthetic translated as a mechanized ethical structure. It is interesting that the French socialist painter Fernand Léger was a lifelong friend of Le Corbusier (Herbert, 2002: 79). Léger's images celebrate humanity in the machine age with pathos and, in his series of large paintings of construction workers, in recognition of the dignity of labour. Léger writes of working-class dance halls as evidence of 'a perfect and complete world that a sick literature prevents us from seeing as it is' (Léger, 1970: 93). Comparison with Le Corbusier's texts highlights a difficulty in that the universal sublime of the machine city and the horizon of progress are idealizations, and as such generalizations; in contrast, the specific, plural experiences of everyday life are open to moments of insight, evidence of a latent utopian consciousness. This is the content of Henri Lefebvre's idea of moments of presence (Shields, 1999: 58–64), proposed in a different way by Walter Benjamin when he sees the phantasmagoria of the arcades as sparking a latent utopian awareness in the observer (1997: 159–60). In a moment of sudden clarity, the world is made new, not in the way Le Corbusier's Obus Plan for Algiers reframes the site with a unity it previously lacked at the cost of erasure, but by seeing it as it is. Whatever is understood is mediated in language, never the unmediated raw of pre-representation, or a mythicized authenticity. Yet there appears a tension between the sky-view, top-down urbanism of the dominant strand in International Modernism, and the street-level, or grassroots, urbanism that Lefebvre references in his concept of lived space that overlays, interrupts and at times reconstructs the uniform – and to Lefebvre Cartesian – domain of conceived space, the space of plans (Lefebvre, 1991).

Perhaps a legacy of Modernist architecture and planning remains this tension, today elaborated in community participation and the design of new public spaces, though problematically in that new spaces do not guarantee the social mixing they are designed to house.

Making new

In Russia, after the 1917 Revolution, the Union of Contemporary Architects favoured modern towers while the Housing Cooperative Movement favoured low-rise housing on the precedent of traditional wooden cottages. The latter was an urban–rural typology to compare with Howard's Garden City, which was known and had been adopted in several pilot projects prior to 1917. Towers were denigrated as 'justified only by super urbanistic aspirations . . . built on the planning theories of Le Corbusier' (Meller, 2001: 126, citing Raputov and Lang, 1991: 805). Infrastructure costs favoured high-rise, however, and Howard's ideas were categorized as bourgeois (while Meller notes that Modernism was read as reacting against the Beaux Arts ideas of the previous century, burdened by bourgeois privilege).

Helen Meller argues that Zlín – a factory town developed by shoe manufacturer Tomás Bat'a in the 1920s – echoes the model of Garden City but that, in continental Europe, 'the only elements which really made the transition . . . were those relating to the achievement of high-quality housing for the workers' (Meller, 2001: 117). After 1918 the class balance had shifted, and Meller continues, 'The garden city was a symbol of a political shift in favour of the lower classes . . . a rallying point for left-wing politicians who eschewed revolution' (Meller, 2001: 119). Henri Sellier, mayor of the French town Suresnes and an admirer of Howard, wrote, for instance, that 'it is possible to guarantee for the working population . . . a form of housing which offers the maximum of material comfort, hygienic conditions which are designed to eliminate the disadvantages of cities, and modes of aesthetic planning' (in Meller, 2001: 121). Bat'a knew of the Garden City and saw Zlín as creating an environment for 'a new civilization' (Meller, 2001: 129) in keeping with his innovative management, in which groups of workers had their own accounts in an internal market, their interactions with other groups translated into a wage policy to encourage personal talents among workers who were 'motivated to work for themselves' (Meller, 2001: 133). In the same spirit, Zlín was professionally designed but, as Meller summarizes:

> Architects worked alongside factory managers, social workers, landscape gardeners and town planners, municipal officials and many other professionals to achieve a modern city. The result was a garden city translated into the idioms of the twentieth century. Zlín became . . . the model of the garden city of the future.
>
> (Meller, 2001: 130–1)

Zlín has single-family, square houses on a grid pattern, a market, exhibition hall, cinema and film studio, schools, a technology college and a range of social spaces. Le Corbusier visited Zlín in 1934 to judge a competition, but his plan for high-rise development there was rejected. Today, an enlightened industrialist might be more concerned with creating sustainable architecture than with the social architectures Bat'a sought to re-engineer. I want to end, then, by looking briefly at BedZED, a suburban housing development in south London designed to enable compact, carbon-neutral living.

BedZED

Beddington Zero Energy Development was designed by Bill Dunster, working with the BioRegional Development Group and engineers Arup, for the housing charity Peabody Trust, in Beddington, near Sutton in south London, a half-hour commute by rail from central London. BedZED is a mixed-use, mixed-tenured site intended for low-impact dwelling, built in low-impact, high-insulation materials on a brownfield site, using renewable energy and harvesting rainwater. It could be called an eco-urban village in its links to ecological living and the urban villages of 1990s compact-city redevelopment. There are 96 housing units, of which 48 (including 14 galleried apartments) were sold as owner-occupier properties, 23 allocated for shared ownership, 10 set aside for key workers and 15 as affordable-rent housing. A typical ground-floor, two-bedroom flat occupies 99 square metres including a south-facing garden. A typical live-work, one-bedroom unit occupies 97 square metres including a sky garden at mezzanine level.

While the average eco-footprint for a British four-person household is 6.19 hectares per person (the land required to support their use of energy and materials), compared to a global average of 2.4 and planetary availability of 1.9, the footprint at BedZED is 4.36 for a conventional lifestyle, but could be reduced to the carbon-neutral target of 1.9 if all residents adopted an ecologically responsible lifestyle (Twinn, 2003: 10; Dunster, 2003: 48–9). The latter would include working as well as living at BedZED, not having a car, recycling 80 per cent of household waste, and eating a low-meat and locally produced fresh-food diet. A conventional lifestyle means owning a car but commuting by public transport, recycling 60 per cent of waste, eating meat and some imported food, and annual holidays by air. Among comparative figures for BedZED are those for electricity – a reduction by more than three-quarters even for a conventional lifestyle compared to the UK average – and water consumption – reduced by one-third (Twinn, 2003: 10). Among design features are the heat-recovery wind cowls on the roof, part of a system engineered to deliver heated fresh air to each bedroom and living room, and extract vitiated air with full heat recovery from all kitchens and bathrooms – a carbon-neutral alternative to air conditioning (or the more usual habit of leaving the central heating on while opening a window to let in some air). While construction costs are higher than for standard volume housing, the benefits in lower energy consumption may compensate, while sourcing materials locally may benefit a local economy. A national precedent was set by extending the legal definition of planning gain to include such environmental benefits (Dunster, 2003: 28).

Dunster claims that the density of BedZED means:

> we could meet almost all the three million or so new homes required by 2016 on existing stocks of brownfield sites . . . [and provide] most new homes with both a garden, a south facing conservatory, and the opportunity to avoid commuting by working on site.
>
> (Dunster, 2003: 8)

High-density use of space and the combination of living and live-work spaces might be as important a measure of the quality of life as low-impact design. Nigel Taylor quotes Anthony Giddens' view that the fragmentation of urban space has meant that 'Place has become phantasmagoric because the structures by means of which it is constituted are no longer locally organized' (Taylor, 2000: 25, citing Giddens, 1990). Giddens himself, in *The Consequences of Modernity*, explains the discontinuity of modern life in terms of its disembeddedness, for instance in the representation of value by money, or of knowledge by technical expertise. This, he argues, relocates trust from human relations to abstract capacities, so that systems are lent a general faith in their efficacy despite an absence for the lay person of any detailed knowledge of how they operate (Giddens, 1990: 21–9). A consequence is that, on the one hand, in personal life, 'trust is always ambivalent, and the possibility of severance is always more or less present. Personal ties can be ruptured, and ties of intimacy returned to the sphere of impersonal contacts', while in public life, on the other hand, demands to be open produce anxiety (Giddens, 1990: 143).

Giddens appears to reformulate the argument for social engineering in terms of a more human-centred model, though in terms also of its absence. Nonetheless, this could be one of several elements in an argument for a compact city in which people not only reduce the impact of their lifestyle by travelling around it less, but also potentially rediscover the sense of proximity that is a foundational quality of city life and produces possibilities for synergy that cannot otherwise occur. This is speculative, but I mention it to raise the issue as to how far BedZED creates the social as well as built architectures that are necessary for sustainability. I do not want to set up a false dualism between BedZED and, say, the community of Tinker's Bubble in Somerset, where a group of 10 people have adopted a non-fossil lifestyle for a small rural settlement (see Chapter Seven). I see no contradiction here, simply a difference of scale and situation (though Tinker's Bubble is also about a right to land). BedZED's urban population can live a non-fossil fuel lifestyle if they choose to, and Dunster advocates using locally sourced components and materials, and recycled materials when available. Wind turbines and cowls, photo-voltaic panels and water conservation are some of the technical means to reduce the footprint in his over-arching vision of a one-planet lifestyle – the level of energy and material use denoting equity in global environmental resource distribution. Dunster imagines zero-energy developments that 'include a farm shop stocking locally produced organic food delivered by solar electric vans' (Dunster, 2003: 48). The performance of BedZED is being monitored and will influence future compact and low-impact developments, and a number of further projects are on the drawing board, such as Sky-ZED, a 35-storey tower to house 300 key workers in co-ownership flats on an inner-city, brownfield site (Dunster, 2003: 148). This would, clearly, rehabilitate the tower block.

The question resurfaces as to how far the design of low-impact dwellings is adequate, not in quality of design but in production of social architectures, to deliver a carbon-neutral city, when well-intentioned modern architects and planners did not deliver a new society by design in the inter- or post-war periods. Is it the other way around, that low-impact living is designed at the grassroots level to be translated into design? Participatory methods were piloted in a limited way by Ralph Erskine at the Byker Wall in Newcastle in the 1960s (Ghirardo, 1996: 148), and by Lucien Kroll in Ecolonia, a housing development at Alphen-aan-den-Rijn near Utrecht, Netherlands (Ghirarado, 1996: 148–9; Barton and Kleiner, 2000: 72). At Ecolonia, 101 housing and office units were built in energy-saving materials, with solar energy and water recycling, in designs produced by nine architects, each establishing different priorities. But Deborah Kleiner cites a post-occupancy survey in which environmental aims were met but significant changes had been made by dwellers, while second-generation owners knew little of the environmental agenda. Would it be fanciful to ask whether a settlement designed and built by dwellers would be socially and culturally more sustainable, if they were provided with a kit of parts for low-impact living? In a 1986 housing project at Admiralstrasse, Luisenstadt, for instance, a concrete frame was used to provide spaces in which untrained residents could erect their own facades, elevations, partitions and balconies (Ghirardo, 1996: 123). I resume the discussion in Chapter Nine on barefoot architecture. Could that be a way forward?

5 Social Utopias

Introduction

In the previous chapter I compared the utopian aspects of Cerdà's plan for Barcelona, Howard's Garden City and Le Corbusier's Modernism. I noted that in the company town of Zlín some of the ideals of Garden City are applied in Modernist forms, and asked how the arguments might be updated in terms of low-impact living. Among issues arising for further consideration were the relation of professional expertise to the tacit knowledge of dwellers, the power relations of philanthropy, and by implication the tension between conventional planning and the need to develop appropriate social architectures. In this chapter I develop the idea of social architecture, looking to departures from the dominant society in the 1960s in the Summer of Love in San Francisco in 1967 and a subsequent growth of intentional communities that can be viewed as doing research and development work for a self-organizing society.

I begin with the Summer of Love, a time of free concerts, squatting, anti-war protest, student occupations, street theatre, communes, marijuana, psychedelic drugs and optimism. I look selectively at groups such as the Mime Troupe and the San Francisco Diggers, noting a parallel history of Situationism in Paris in the 1960s in which, too, a society based on productivity is challenged by non-productive uses of time. In the next section of the chapter I chart a move to intentional communities in North America, often in remote sites, where alternative lifestyles could be pursued away from media attention. This was not a return to a Rousseauesque vision of a natural past but aligned with alternative values found, in some cases, in indigenous cultures. I question the appropriation of tribalism, and ask how it compares to the post-colonial search for negritude.

If you go to San Francisco . . .

In 'Slouching Towards Bethlehem', Joan Didion describes the Haight-Ashbury district of San Francisco – the District – in 1967, the year of the Summer of Love, as where the children of American middle-class families went when they ran away from home. This is literal and metaphorical. For the young people, the post-war generation, it was an escape from the consumerist suburban lifestyle their parents saw as the optimum social state. Metaphorically, it was an escape from America

– the consumerist republic of the affluent society where Liberty had become atrophied as consumer choice, and affluence was excess. Didion writes, 'All that seemed clear was that at some point we had aborted ourselves and butchered the job . . . San Francisco was where the social haemorrhaging was showing up . . . where the missing children were gathering and calling themselves "hippies"' (Didion, 2001: 72). The withdrawal was called the Movement. Those who constituted it were mainly but not exclusively white, middle-class and in many cases college-educated (Doyle, 2002: 81). But the withdrawal from the dominant society was not an escape into personal fantasy, and while it was personal for those taking part, and reliant to an extent on psychedelic drugs, it was social in as much as the Movement was a counter-cultural formation. The interest of Didion's account is that she brings out the politicized aspect of the counter-culture that media images of flower children denied.

The Movement, then, was a social re-formation contextualized by student and anti-war protest, civil rights marches and black power, solidarity with struggles for national liberation in the non-affluent world, and efforts to define a New Left, as well as economic boom. It seems false to separate the everyday from the politicized aspects of the counter-culture, and more helpful to see the everyday life of the Movement as a refusal of the system that produced the Vietnam war (just as the European Dada movement had been a rejection by artists and writers of the values that had produced the 1914 outbreak of war). If the Summer of Love is a retrospective term for a lifestyle of new music and the use of marijuana and psychedelic drugs, this was revolt as indicated in the slogan 'Make Love Not War'. It was, too, in many aspects of everyday life in the District, an adoption of non-productive uses of time as resistance in a society geared to productivity (see Illich, 1973, 1996).

This might be read as coincidentally parallel to the non-productive use of time as an art form in the urban drift (*dérive*) of the Situationists in Paris. There, from the early 1960s, artists had moved aimlessly (non-teleologically) through the city rediscovering traces of its radical past and finding new insights in chance encounters (see Pinder, 2005: 126–59; Sadler, 1999: 29–30, 77–8). This non-end-driven drifting, partly art, partly psycho-geography, partly ordinary life (with wine), differs in character as well as means from the use of sabotage by the Weathermen in North America and the Red Army Faction in Germany – where Cold War tensions were at their highest and the student movement strongest (Varon, 2004: 38) – and differs equally from the political campaigning of civil rights marches in the United States in the early 1960s. Yet if sabotage and civil rights are overtly political or insurrectionist, the aimlessness of the drift is its revolutionary quality in a refusal of the teleological instrumentalism and productivism of the dominant society. As David Pinder writes of the *dérive*, 'Accounts . . . suggest slowness and a sense of drifting with currents . . . fugitive movement, willed drive and an intense sensation of the passage itself' (Pinder, 2005: 151). Pinder cites Kristin Ross that the *dérive* resonates with a refusal of the idea of work as toil in the writing of Rimbaud and Laforgue, and in the oppositional culture of the Paris Commune of 1871, and quotes Ross that laziness becomes 'the impossible liberty of having exempted oneself from the organization of work in a society that expropriates the very body of the worker' (Ross, 1988: 60, in Pinder, 2005: 152).

If by analogy the aimlessness of Situationism (using the term aimless to play on its meaning as refusal of an objective as well as drifting) is revolt, so the counter-culture in which life and art were at times merged in San Francisco in 1967 might be regarded as revolt. The contexts are different, and I draw on Situationism only to open the question of how to read the Summer of Love. The Situationists were an artistic avant-garde informed by Henri Lefebvre's theories of liberation (Pinder, 2005: 131–44; Shields, 1999: 58–64) while the Movement was disparate, at times contradictory in its practices, more attuned to personal liberation than the evolution of a new geography of resistance. Yet perhaps the Movement enacted (at the level of a mass public within the District) an alternative way of living that could be called a resistance within everyday life. Perhaps, too, its general libidinization of social formation could be compared to, or seen as a remote adaptation of, Charles Fourier's idea of a libidinal society (though Fourier retains the centrality of work as the cooperative basis of this libidinization, as explained in Chapter Three). But I was not in San Francisco, as an art student in London from 1967 to 1971, and my interpretation is based on reading other people's accounts. I want simply to speculate that the Summer of Love was not only about wearing flowers in the hair, although flower power, even as a linguistic gesture, questions the violence on which the dominant society depends. Herbert Marcuse states in his opening remarks to the Dialectics of Liberation Congress at the Roundhouse in London in July 1967, after expressing his happiness at seeing so many flowers present: 'flowers, by themselves, have no power whatsoever, other than the power of men and women who protect them and take care of them against aggression and destruction' (Marcuse, 1968a: 175). The content of his talk, nonetheless, was liberation from the same affluent society from which the flower children ran away to San Francisco.

Jeremy Varon (2004: 37–8) sees Marcuse, who spoke frequently at student meetings in 1967 in Europe and North America, as one of the Movement's theorists, beside Karl Marx. Varon identifies the Black Panthers as the Movement's vanguard, the Yippies as its tricksters and the San Francisco Diggers as its visionaries. He notes the prominence of underground magazines in the counter-culture, in which student radicals 'debated ideology, announced demonstrations, denounced the police, reviewed albums, concerts, books, and plays, and most broadly, shared their vision of themselves and the world', and summarizes that the New Left 'imagined itself a liberating agent of history' (Varon, 2004: 38). This may have applied particularly to the student movement, Students for a Democratic Society (SDS) in the United States (see Varon, 2004: 20–30, 44–54), and even more so to student and other radical groups in Europe who sought to go beyond the conventional politics of representation, using tactics such as small-group direct action and mass demonstrations. For such groups an immediate political issue was opposition to the Vietnam war, but resistance was theorized, too. Parallel to the non-violent action of student movements was, Varon narrates, the emergence of revolutionary shock-troops in the Weathermen and the Red Army Faction, prepared to use violence and sabotage to combat the violent state (or military-industrial complex).

78 Practices

In the District, the tools of resistance were the non-violent means of hanging out, making love and turning on. Didion's account of life in Haight-Ashbury also mentions eating macrobiotic food, light shows, music and street performance. Didion is immersed, as a journalist, in this ephemeral side of life in the District but, as mentioned above, she rejects the media image of the counter-culture as depoliticized. While, that is, images of flower children sitting around and getting high, indulging in what could be dismissed as a teenage rebellion on a countrywide scale, represented no real security threat, the activities of the San Francisco Mime Troupe, the San Francisco Diggers, the Merry Pranksters, the Hog Farmers and the Motherfuckers – the latter returning to the society that donated it the abusive name they received from its law enforcement officers – which are recounted in more detailed histories of the Summer of Love, represent a politicized dimension, using street theatre to merge art and everyday life and provoke the dominant society into revealing its real character. In this aspect of the Movement there were links between dropping out and student protest, and a general opposition to the war state. Alberto Buenfil reads the Gathering of the Tribes for a Human Be-In, in Golden Gate Park in the summer of 1967, as an effort to link all the main currents of the youth movement, 'the groups that formed the "New Left", . . . that opposed the war in Vietnam and . . . Third World revolutions . . . [and] who identified more with the hippies, communal extended families, new tribes, flower children, and the love generation' (Buenfil, 1991: 52). This demonstrates a historical connection between flower power and programmatic groups but, again, I want to avoid the impression that there was, on one side, a mass dropping-out and, on the other, a political campaign. For the New Left and the SDS, perhaps freedom was a project with aims and tactics, while for those outside such movements who had gone to San Francisco in 1967, refusal of the affluent society was implicit in a new way of living. Somewhere between these positions was street theatre, articulating critical messages in disciplined but open-ended and adaptive ways, and more challenging because it had no script, no solution to the problem of America.

The currents of tacit and programmatic resistance are evident in the following two passages from Didion's account of her time in the District:

> It is a pretty nice evening and nothing much happening . . . The Warehouse, which is where . . . a floating number of other people live, is not actually a warehouse but the garage of a condemned hotel. The Warehouse was conceived as total theatre, a continual happening, and I always feel good there. . . . Somebody is usually doing something interesting, like working on a light show, and there are a lot of interesting things around, like an old Chevrolet touring car which is used as a bed and a vast American flag fluttering up in the shadows and an over-stuffed chair suspended like a swing from the rafters, the point of that being that it gives you a sensory deprivation high.
>
> (Didion, 2001: 81)

and:

> the political potential of what was then called the movement was just becoming clear. It had always been clear to the revolutionary core of the Diggers, whose

every guerrilla talent was now bent towards open confrontation and the creation of a summer emergency, and it was clear to many of the straight doctors and priests and sociologists who had occasion to work in the District . . . One did not have to be a political analyst to see it; the boys in the rock groups saw it . . . 'In the Park there are always twenty or thirty people below the stand,' one of the Dead complained to me. 'Ready to take the crowd on some militant trip' . . . the peculiar beauty of this political potential, as far as the activists were concerned, was that it remained not clear at all to most of the inhabitants of the District . . . Nor was it clear to the press, which at varying levels of competence continued to report 'the hippie phenomenon' as an extended panty raid; an artistic avant-garde . . . or a thoughtful protest, not unlike joining the Peace Corps, against the culture which had produced Saran-Wrap and the Vietnam War.

(Didion, 2001: 103–4)

Didion describes different dimensions of life in the District in separate passages, but her argument is that the generation educated between 1945 and 1967 saw the values by which they had been educated as rendered vacuous by the affluent society. Elsewhere, in a similar axis of art, life and resistance, there were student sit-ins and demonstrations in reaction to the unresponsiveness of universities to issues such as the Vietnam war, and a breakdown in the credibility of authority (Searle, 1972: 151–7). There were also new, autonomous Free Universities as an alternative, non-hierarchical structure. Marcuse and Joseph Beuys lectured at the Free University in Berlin in 1967; and Theodore Roszak notes an Anti-university of London offering courses in anti-cultures, anti-families, and other such negatively put subjects, eventually refusing the teacher–student power-relation by scrapping formal teaching: 'nobody any longer had anything to teach the young; they would make up their own education from scratch' (Roszak, 1995: 46). That was in 1968, the year in which art students occupied Hornsey School of Art in London and free concerts took place in Parliament Hill Fields with bands such as Jefferson Airplane, and when anti-war protest increased as the prospect of a US withdrawal from Vietnam seemed possible. And 1968 was also the year in which an alliance of workers and students in Paris came, arguably, within sight of bringing down the French state.

Much in French theory after 1968 is a rationalization of the failure to achieve that change; yet as in San Francisco the revolution was not only of the traditional kind made in barricades, but was constituted as much in new social formations and ways of living within a radical milieu. Julia Kristeva recalls her experiences in Paris:

the liberation of social behaviour was an essential experience of '68. Group sex, hashish, etc., were experienced as a revolt against bourgeois morality and family values. . . . This movement can only be described as political because it began by striking savagely at the heart of the traditional conception of love. . . . We often had the impression that as soon as something new was on offer, it became an institution. As far as sex is concerned, most didn't have this impression.

(Kristeva, 2002: 18–19)

This might suggest Charles Fourier's idea of sexual liberation in the phalanstery, except that Kristeva refers to the actualities of living in a cosmopolitan milieu. In the District, meanwhile, the hippie (Haight-Ashbury independent property) culture had taken an equally resistant but different form in 1967. James Farrell writes:

> The hippies of the Haight differentiated themselves from mainstream culture by their drugs and music, but also by their hair and dress and decorum. Like civil rights workers who put their bodies on the line, countercultural activists drew a line with their bodies. When men let their hair grow . . . their long hair . . . symbolized countercultural identity and defiance of the culture of conformity. . . . Freeing themselves from the fashion world, many wore hand-me-downs and Army surplus . . . Men and women adorned themselves with flowers and with crafted beads and colourful baubles. . . . By appearance and behaviour they declared themselves actors (and sometimes activists) in a new cultural drama.
>
> (Farrell, 1997: 219)

Frances Fitzgerald writes similarly that the hippies aimed to 'disarticulate the society and the intellectual frames they had grown up in', and that 'everyone had a right to do his or her own thing. If they were gay, that was their thing' (Fitzgerald, 1987: 43). For others it was psychedelic drugs, as in Timothy Leary's call to drop out and turn on. Leary uses a biological metaphor for a process of free expansion followed by the reactive contraction of institutional forms:

> The expanding gaseous cloud whirls into temporary patterned structures . . . watery, electro-biochemical globules cluster into cells . . . The free, expansive vision is moulded into the institutional . . . [and then] a new cortical upheaval, an explosive, often ecstatic or prophetic revelation.
>
> (Leary, 1970: 55)

I have reservations about biological metaphors that universalize the conditions they describe, taking them out of history; but Leary's image may denote the evolution of an individual consciousness and a social movement, or both.

Revolution by other means?

Alongside dropping out and turning on was the street theatre of the San Francisco Mime Troupe, a fusion of highly disciplined theatre and free, open-ended articulation of the social moment. The Mime Troupe had given street performances since 1961, and in 1964 had established a Free University in their studio in the Mission district, linked to the New School for Social Research. Then, in 1965, a performance of Giordano Bruno's *Il Candelaio* was labelled obscene by the city's Park and Recreation Commission. From then on, initially through street protest, the Troupe adopted more confrontational tactics. Among the outcomes was a parody of the conventional blacked-up minstrel show, *A Minstrel Show or: Civil Rights in a*

Cracker Barrel (Doyle, 2002). Drawing on Che Guevara's idea that guerrilla groups survive only through the support of the oppressed they seek to liberate, the group adopted the idea of guerrilla theatre. Michael Doyle summarizes: 'the guerrilla cadre provided a model worth emulating . . . Both [cadre and theatre group] were small, highly disciplined groups . . . Motivated . . . to do battle against enormous odds' (Doyle, 2002: 74). Didion saw Mime Troupers improvising in Golden Gate Park, forming a circle round

> a Negro . . . jabbing the Negro with the nightsticks . . . 'Just beginning to get annoyed, are you?' one of the Mime Troupers says, 'Don't you think it's about time?' . . . It went on for a while. But on that particular Sunday . . . everyone was too high.
>
> (Didion, 2001: 107–8)

Didion sees the black participants as uncomprehending, there for the music and rejecting the idea that black music had been stolen. But Doyle points out that civil rights activists were encouraged to audition for parts with the Mime Troupe, and reads *A Minstrel Show* – not what Didion saw – as adapting a popular theatrical form, as they had in their use of the Commedia dell'Arte, 'to explore a series of wide-ranging contentious topics, in this case selected from more than a century of American racial discourse' (Doyle, 2002: 75).

In 1966 a group of 20 or so members of the Mime Troupe broke away to form the San Francisco Diggers. Doyle writes of their 'vision of the total transformation of social and economic relations' and dedication to bringing about the New Jerusalem by peaceful means through pamphlets and 'exemplary actions' (Doyle, 2002: 79). The Diggers used the theatrical vocabulary of the Mime Troupe, developing a fusion of life and art in life-acting, using participatory role-play – probably what Didion saw – to act out scenarios of social relations. Life-acting, they argued, produced a new consciousness by interrupting the routines of daily lives. A high, yellow, wooden frame (the Frame of Reference) was built at the intersection of Haight Street and Ashbury Street as a site for puppet theatre, through which participants in life-acting walked in a symbolic entry into an expanded consciousness. Participants were guided in the game of Intersections, in which they crossed the streets 'in a way that traced as many different kinds of polygons as possible . . . to impede vehicle traffic . . .' (Doyle, 2002: 85; see also Farrell, 1997: 220). It also impeded the flow of tourists in cars for whom the district had become a new day-trip destination. The Diggers distributed free food, opened a free store (the Free Frame of Reference), claimed free housing in squats, set up a free medical clinic and organized free concerts by Jefferson Airplane, the Grateful Dead, and Country Joe and the Fish. They advocated free use of psychedelic drugs, and for David Farber this is key to their identity, though he recognizes that 'acid was not the only fuel driving Digger energies' (Farber, 2002: 30). He continues that they aimed to de-school individuals, and by that route to de-school society, overcoming the limits to consciousness in part but not only through psychedelic drugs, thereby disconnecting from the value systems of capitalism and the Cold War. The Diggers drew on the free time of the young people who escaped to the District, and on the surpluses of the affluent society recycled as free goods, to create a free economy standing for a free society.

Other radical groups were similarly located in the overlapping terrains of psychedelia, direct action and withdrawal to an alternative lifestyle (later in intentional communities), among them the Merry Pranksters and Hog Farmers in California, and the Motherfuckers (Buenfil, 1991: 49–61). There was a New York Diggers group as well, and the Yippies (Youth International Party – Doyle, 2002: 85–97). Buenfil aligns them all to 'anarchistic movements of the Middle Ages and their messianic communities' (Buenfil, 1991: 58), citing Murray Bookchin, Norman Brown, Paul Goodman and Noam Chomsky, as well as Marcuse, as their theorists. To exemplify the character of the Motherfuckers he cites the following passage from the counter-culture magazine *Black Mask*, founded by artists Ben and Janice Morea and Allen Hoffman in 1966:

> A new spirit is rising. Like the streets of Watts, we burn with revolution. We assault your Gods. We sing of your death. *Destroy* the museums . . . our struggle cannot be hung on walls. Let the past fall under the blows of revolt. The guerrilla, the blacks, the men of the future, we are all at your heels. Goddam your culture, your science, your art. What purpose do they serve? Your mass murder cannot be concealed. The industrialist, the banker, the bourgeoisie, with their unlimited pretence and vulgarity, continue to stockpile art while they slaughter humanity. Your lie has failed. The world is rising against your oppression. There are men at the gates seeking a new world. The machine, the rocket, the conquering of space and time, these are the seeds of the future which, freed from your barbarism, will carry us forward. We are ready. *Let the struggle begin!*
>
> (Ben Morea and Ron Hahne in *Black Mask*, 1 November, 1966, quoted in Buenfil, 1991: 59)

The vocabulary echoes Futurist manifestoes of the 1910s, not least in using a universal masculine. Setting that aside, the text locates the struggle in a terrain of museums, or as counter-culture in the tracks of Modernist avant-gardes attacking bourgeois society via its institutions. In fact, those Futurists who survived the 1914–18 war, which they saw as a necessary purging of society, supported Mussolini in the 1920s.

Setting that aside as well, the above text is implicitly performative, like happenings with a purpose. The time of contemplation is over; that of action begins. Similarly performative is the work, described by Doyle, of the New York Diggers:

> The group arranged for a tour of the New York Stock Exchange (NYSE) under the auspices of ESSO (the East Side Service Organisation, a hip social services agency; the fact that this acronym was better known as the name of a giant oil corporation is probably what gained them entrée to the NYSE). Once they had been escorted into the visitors' gallery . . . they produced fistfuls of dollar

bills and flung them from the balcony onto the floor below. All bidding stopped as traders impulsively switched . . . to an atavistic frenzy, scrambling to grab what they could from the shower of cash. They then began to berate the Diggers.

(Doyle, 2002: 86–7)

These groups, spreading from the West to the East coast, denote a phase of the counter-culture engaging actively and at times (non-violently) aggressively with the dominant culture. They also began to develop as alternative social formations, evolving their own structures as self-organizing, autonomous entities. Farrell summarizes:

> The counterculture combined a comprehensive cultural critique with an eclectic collection of oppositional cultural institutions. The critique challenged American culture's dominant myths – the work ethic, utilitarian individualism, repressive sexuality, Cartesian rationality, technocratic scientism, denominational religion, industrial capitalism, lifestyle suburbanism, and compulsive consumerism – and offered intellectual and institutional alternatives.
>
> (Farrell, 1997: 204)

This differs from the ethos of the Diggers of the 1640s (see Chapter Two) for whom work was a means to build a new society, as in cultivating the repossessed land, and is closer to the Situationists' retrieval of a graffiti from the 1950s stating '*ne travaille jamais*' (never work) on a wall in rue de Seine in Saint-Germain-des-Prés, Paris (Pinder, 2005: 144, fig. 5.6). Yet there is a tension in English radicalism in the 1640s between religiosity and rationality, and between individualism (or individual wealth) and collectivism, suggesting that a comparison between activism in the 1640s and the 1960s might draw out insights applicable to both, despite the obvious anachronism. Key to this is whether the revolution is imminent – soon to come, realized through incremental reform and instrumental action – or immanent – immediate and all-pervasive, non-instrumental and non-teleological because the means are seen as the ends themselves, the enactment of freedom as a free society in the present. One way in which an immanent revolution occurred was in the move to intentional communities, beginning as a withdrawal from the withdrawal that had been the Summer of Love.

After the Summer of Love

Farrell reports that by the mid-1970s there were around 7,000 communes in North America, housing several million people, most of them in cities. Examples included New Life in Philadelphia, a Quaker social action community; Bread and Roses Collective in Boston, a women's refuge; 95th Street Collective, New York, a gay community; and the socialist Kate Richards O'Hare Collective in Ithaca, near Cornell University (Farrell, 1997: 224; see also Miller, 2002: 337–8, 341–7). There were spiritual groups such as the Lama Foundation (Miller, 2002: 336; McLaughlin and Davidson, 1985: 282–5), the Chinook Community, the Abode, and the Providence Zen Centre (McLaughlin and Davidson, 1985: 260–4, 273–6, 277–82), exploring a range of mainly non-Western traditions, closeness to nature and meditation practices. Others found remote sites where they could develop nonconformist and non-violent lifestyles. Black Bear Ranch, for example, founded by a breakaway group of Motherfuckers, is in a remote, mountainous area of California. Buenfil cites an interview with Peter Berg and Judy Goldhaft, theatre activists who stayed at Black Bear for a few months:

> A lot of us Diggers went and lived at Black Bear Ranch for a while . . . it was a radical commune, maybe the most radical in the '60s. . . . the only place I've heard of where 50 adults decided to live in one room for the winter and have multiple sexual relationships . . . a tremendous emotional wreckage. . . . But it was also at Black Bear that we started to develop a sense of place. We started to become conscious about the water of the creek.
>
> (Buenfil, 1991: 60)

Despite the emotional drain, Black Bear survived, becoming marginally integrated into the local economy when some members became part-time forest firefighters. Timothy Miller writes, 'Black Bear's great isolation caused it to become one of the most self-sufficient communes' (Miller, 2002: 334, see also Black Bear Ranch, 1974; Monkerud, 2000). Members of the commune learned to live from the land, hunt, deal with medical problems, and run a saw-mill and hydroelectric power source. Miller cites another case, New Buffalo near Taos in New Mexico (an established node of alternative living and adobe architecture), where members lived in tepees and built a common house in adobe, seeking to be near Native American culture as well as the Earth. As Miller says, 'it was never an easy life; agriculture was difficult in the arid climate . . . and the cold winters challenged all but the hardiest' (Miller, 2002: 335; see also Fairfield, 1972: 187–97). In retrospect some of this might seem a too-easy appropriation of non-white pre-industrial lifestyles. I take up that critique below in discussion of the emerging alternative society's attraction to tribalism. Perhaps, too, despite the hardness of living on the land for people raised in an industrialized society, it was therapeutic.

In some cases the departure was a means to move away from hard-drug culture. The Farm, for example, was founded by a group of young people who left San Francisco in a fleet of buses and trailers, with philosophy lecturer Stephen Gaskin. The first trip was nominally a speaking tour by Gaskin but produced the idea of a long-term departure from the city. In 1971 the group moved to Summertown in rural Tennessee. Three hundred and twenty settlers built shelters, mainly log cabins, and began to cultivate the land, aiming to live self-sufficiently and adopting a vegan lifestyle. After four years the Farm had become more or less self-sufficient, and the population increased to over 1,000. The settlement included a school, grocery stores, greenhouses, machine shops, a publishing company and a car repair workshop. Dwellers agreed to live on $1 per day, distributed produce equally and shared cooking and laundry facilities; many lived in common houses and practised home-birthing methods. Miller writes that 'They were forthcoming about the psychedelic dimension of their spirituality' and that the community learned from the low-tech lifestyle of their Amish neighbours (Miller, 2002: 337; see also Traugot, 1994). The Farm attracted a large number of visitors, and its population reached 1,500 by 1983, when it underwent a financial crisis. Walter and Dorothy Schwartz, who visited the Farm in the 1990s, write: 'Debts could no longer be paid. What happened is still argued about' (Schwartz and Schwartz, 1998: 271). A decision was taken by some means that only self-sufficient members could stay, paying dues to the Farm for land and services. Schwartz and Schwartz add that despite this the ideals of the Farm – its frugality and sense of community – survive, and that some members again pool their earnings. But the Changeover – the events of 1983 – implies a collision of personal and social dimensions, and the difficulty of building the immanent society in a project requiring the management of resources, a balance of diverse interests and maintenance of a unifying vision that had initially been provided by Gaskin in a teacher–student relationship in which he was the determinant partner, as in the group's sexual ethos. Timothy Miller concludes:

> Saving the world headed the Farm's agenda, and the commune soon found itself running a wide range of charitable projects. Always family-oriented and opposed to abortion, the Farm developed a circle of midwives who ran a birthing clinic . . . A foundation called Plenty undertook projects that ranged from Third World agricultural education to an ambulance service in the Bronx. For more than a decade the community was a bright beacon . . . [After 1983] the community went through a radical depopulation, but after a time it regained some stability.
>
> (Miller, 2002: 337)

This illustrates some of the difficulties. David Miller writes that some structures of community are 'antithetical to other equally important, socialist ideals' (Miller, 1992: 85–6), and the histories of 1960s communes and intentional communities indicate a need for ways to address the organization of work outside the post-scarcity conditions of San Francisco in 1967. But this extends to all areas of daily life, from washing up to sex and growing food, and involves continuous interaction. That it is viable to work through the difficulties is demonstrated by the extent to which groups which are in effect elements of an alternative society have done so in intentional communities. To give some data, McLaughlin and Davidson list 117 communities in *Builders of the Dawn* (1985: 348–9). Of them, 103 are in the United States, three in (each of) Canada, India and Britain, and one in each of New Zealand, South Africa, France, Denmark and Israel – a by no means comprehensive list outside North America. Of those in the United States, a majority are in either California and Oregon or in New England. A directory of European intentional communities and ecovillages for the year 2000 has 326 entries in 24 countries, of which, in overlapping categories, 217 see themselves as ecological, 180 autonomous, 159 cultivate food or forests, 138 are organic and 112 vegetarian,

85 are self-sufficient and 16 practise free love (*Eurotopia*, 2000: 7). A directory for the United Kingdom lists 89 communities in 2004, from the long-established (such as Findhorn in Scotland, founded in 1962) to the emergent, including spiritual, ecological and urban groups (Diggers and Dreamers, 2004). Many such communities offer programmes of education and personal growth. For example, Lebensgarten in Germany, founded in 1985 and housed in a group of ex-Nazi armsworkers' buildings, states: 'Our seminar work stands at the centre of the community. Here the experiences which we have made ourselves can be passed on to others. Personal development, ecology, health, meditation, and conflict resolution are important topics' (*Eurotopia*, 2000: 138).

This establishes that a plural alternative society exists. Accounts of many of the sites show that groups (mostly from a dozen to a hundred people, often of all ages) are doing the work of alternative living, dealing with conflicts among members and negotiating a relation to the wider society while refusing the violence and oppression of the dominant society. I end this section of the chapter by citing Judson Jerome's *Families of Eden* (undated), quoted by McLaughlin and Davidson:

> Those who worked hardest and most consistently were inclined to interpret the sporadic efforts of the others as lack of commitment . . . Those who do more than their share resent those who do little. The latter accuse the former of being judgmental . . .
>
> The games of ego can be played with monetary or sexual counters, and also those of labor. One person's exertion can be interpreted as pressure on others to exert equally. Thus one can develop an ego-stake in *not* working, a resistance to being manipulated . . . The right to freedom of choice is perpetually tested.
>
> (McLaughlin and Davidson, 1985: 69–70)

Tribes

I move now to the tribalism of the counter-culture. Buenfil describes the Hog Farm in the San Fernando Valley, California as 'one of the best known of the extended families, the new tribes, the experimental clown armies of the last thirty years' (Buenfil, 1991: 53). The Hog Farmers used street theatre, music and light shows, and were mobile. With the Merry Pranksters and Motherfuckers they joined antiwar demonstrations, engaged with ecology and called for the legalization of marijuana. The three groups met in Santa Fe in 1969, celebrating the solstice with a race in their buses. The Hog Farmers went to London (where they bought a red bus) and toured Europe and South Asia. In Germany they bought another bus, naming it the Rainbow Repair-shop. Meanwhile, Peter Berg and Judy Goldhaft set up Planet Drum as an organization for bioregional and ecological projects, and in 1968 the Death of Hippie ceremony organized by the San Francisco Diggers marked the close of an era. Buenfil comments, 'it was a sign that something had started to change . . . The ceremony meant that a change of consciousness was necessary in order to move on to the next step. . . . it was time to start confronting the system in a more aggressive way' (Buenfil, 1991: 61). Perhaps the moment of

1967 could not have lasted anyway except in institutionalized forms, as when cafés in San Francisco gave free coffee to hippies to sit at their tables as a way to attract tourists; but the above indicates a resumption of instrumental tactics as well as a more aggressive stance. The alternative, still possibly a way of life rather than a campaign, was nomadism, the construction of a second counter-culture in which refusal extended to the city. Garrick Beck describes the Rainbow Gatherings which began in 1970: 'back-to-the-landers, food co-ops, Eastern mystics, and worker collectives' went separate ways while 'Marxist-Maoists were theorizing in progressively smaller circles' and 'a number of tribally oriented groups' met at an arts festival in the Pacific Northwest (Beck, 1991: 68). He continues:

> We envisioned and planned all at once. The gathering was to be . . . publicized largely by word of mouth. As we talked with others, every so often someone would light up in response to the idea, as though they themselves had independently seen or desired the event to be. These people spread the word. We saw ourselves not so much as innovators as we did messengers called upon to revive an old form of human congregation and communion.
> (Beck, 1991: 68)

The first gathering was in Colorado in July 1972, the name Rainbow Family reflecting the ethnic diversity of those who took part, including invited Native Americans. Efforts to gain official permission for the gathering were met with indifference, then a closing of all campgrounds in the area and call-up of the National Guard. Around 2,000 people arrived, nonetheless: 'Faced with no realistic alternative, the sheriff's department let the swarms of singing, praying, peaceful marchers pass on by. The column continued for the next three days . . . up the next six miles to Strawberry Lake' (Beck, 1991: 70). The form of the gathering set a pattern for annual events. In the centre of the site is a place for silent meditation. Workshops cover health, diet, yoga, massage, weaving and other crafts. A Centre for Alternative Living Medicine (CALM) provides free health care. Each sector of the camp has a communal kitchen where performances by poets, storytellers, dancers and musicians take place at night. A group of security workers receives prior training in non-violent persuasion, peer-group pressure and reintegrating the alienated. A decision-making body, the Tribal Council, is open to all, operating both as a whole and in break-out groups. Beck sees it as 'not only an effective means of self-organization but also . . . of preventing political decay' (Beck, 1991: 75). There are tepees and sweat-lodges, but 'while tipi life and traditional ways are evident . . . so are the hi-tech wonders of the future' (ibid.) – wind-driven communication systems, solar heaters and silicon lighting.

A sign at the 1990 Rainbow Gathering in Minnesota read, 'It is said that when the Earth is weeping, and the animals are dying, a tribe of people will come who care, they will be called the Warriors of the Rainbow' (Buenfil, 1991: plate 57). Buenfil writes, too, of 'the tribalism of the rainbow' beginning in the beat generation's travels to North Africa and Asia. He sees the poet Gary Snyder as bringing together studies of Zen and Native American cultures after a long-term association with libertarian movements, and quotes a passage of Snyder's that ends: 'Men, women and children – all of whom together hope to follow the timeless path of love and wisdom, in affectionate company with the sky, winds, clouds, trees, waters, animals and grasses – this is the Tribe' (Snyder, 1957: 116, quoted in Buenfil, 1991: 42). Buenfil adds Snyder's description of the 1967 Human Be-In: 'the new aboriginals gathered . . . some of them with their own flags and banners. Those were the real tribes' (Snyder, 1957: 93, quoted in Buenfil, 1991: 43). But most were white and middle class, and their attraction to tribalism may require explanation.

In North America, tribal myths were perpetuated in films and television programmes featuring Native Americans, in museum holdings and in a growing literature of indigenous American writing through which indigenous cultures were perceived as caring for the wild and being frugal. But was this white adoption of native cultures, often generalized as a single entity, an uninformed, romantic appropriation of another people's culture? I have a similar difficulty with art critic Suzi Gablik's use of the term shamanism to describe non-gallery art in the 1980s in *The Reenchantment of Art* (1991). The cultural lineage is not unclear, however. First, African masks represented a radical-other cultural form rupturing the perceptual codes of the European tradition for white Modernists in the 1910s; indigenous American culture, interpreted as tribalism, may have similarly been a catalyst for the counter-cultural groups of the 1970s. If so, Daniel Miller's critique of primitivism is appropriate: 'Primitivism stands for that aspect of the Romantic movement which is based on the assumption that there exists a form of humanity which is integral, is cohesive, and works as a totality' (Miller, 1991: 55; see also Lloyd, 1991). Miller adds that this is defined as not-the-present, as Other, or as childlike. Second, the 1950s idea that artists plumb depths of psyche others fear to tread – in the work of artists such as Jackson Pollock and Mark Rothko (Kuspit, 1993) – leads when that fails to a turning away from previous avant-gardes in a search for a new authenticity of direct experience. This is mythicized because everything in language, in any cultural medium, is mediated, and thus representation in some form and never direct experience (which is pre-verbal, just as Miller's cohesive humanity is pre-natal). Loss of a direct connection to experience, like the perceived loss of unity, produces compensatory attractions to indigenous American culture as the location of a projected authenticity or wholeness despite the fact that (or because) little is known of the cultural forms involved by those who appropriate them. Similarly, Gablik begins from the position that Modernism failed to retain an authentic voice of dissent, turning to a purely formal language, and perhaps it is her sense of loss that leads her to look to a borrowed vocabulary to describe, and project onto, such art as she finds meaningful.

The quest for authenticity in a society trading on the artificiality of consumerism raises the question as to how wants are produced or differentiated from needs (Papanek, 1984), or whether the affluence of a consumer society produces new needs when that for sustenance is satisfied (Lodziak, 2002). I have to leave the sociological argument there, however, to take up the argument instead that the quest for authenticity that is perpetuated through appropriation of other people's cultures differs at root from the development by black poets Aimé Césaire and Léopold Senghor of the concept of negritude in the 1930s (Césaire, 1997; Senghor, 1998), and from the sense that, for Frantz Fanon (1967, 1986), blackness is always polyvalent (see Venn, 2000: 97). The example of post-colonialism was important to the European Left, and Varon notes Jean-Paul Sartre's preface to Frantz Fanon's *The Wretched of the Earth* (1967): 'We only become what we are by the radical and deep-seated refusal of that which others have made of us' (Varon, 2004: 88). It is straightforward to see this as exemplifying post-colonial cultures, but Varon also cites Ralph Larkin and Daniel Foss looking back on the North American New Left, who say, 'to be "real" for New Leftists meant "being what one becomes upon rejection of the conventions" learned through one's mainstream socialization' (Varon, 2004: 88, citing Larkin and Foss, 1988). A difference between the white New Left and black power activists in North America and those who, like Césaire, Senghor and Fanon, sought to extricate themselves from the cultural webs of colonialism, was the sense of loss, not of a perceived authenticity but of a material culture and a sense of self. Couze Venn writes, for instance, that 'Fanon's account of the sense of annihilation he experienced, and the Negritude writings to which he turned in search of another speech, express a multiplicity of losses: of identity and self-esteem, of history and rootedness, of face and agency' (Venn, 2000: 105). The runaway children in San Francisco did not feel that, and neither did the New Left (inasmuch as I can say from secondary sources).

The comparison reveals a rupture of the liberal-humanist subject, the unified self of Enlightenment discourse who was the slave-owner and colonial administrator. John Docker argues that if white poetry resolves social tensions in 'a metaphysical idea of human freedom which is superior to history', post-colonial literatures open 'questions of community, of ethnic and national identity, of the cultural effects of industrialization and urbanization, and of the continuity or discontinuity of traditions' (Docker, 1995: 444). In place of a false idea of unity, then, is the idea of a contingent, negotiated and hence always unfinished equilibrium in the interaction of specific human agencies in specific conditions. The choice is whether to work with that or to escape to a therapeutic idea of otherness in tribalism. Philip Deloria writes of childhood visits to a commune in the Pacific Northwest in 1971 where 'a reassembled "family" of non-Indians [*sic*] . . . eked out a living making Aleut-modelled soapstone carvings' in a group of tepees (Deloria, 2002: 159). It was probably Black Bear Ranch (see above). He notes that tipis come from plains cultures while the vernacular of the Pacific Northwest is the cedar cabin, and remarks that in a crisis, 'white Americans . . . inevitably turned to Indians . . . [but] remained largely disconnected from Indian people' (Deloria, 2002: 161, 163). Sitting Bull and Geronimo were appropriated by the anti-war movement, for instance, when many Native Americans were soldiers in Vietnam. Meanwhile indigenous American consciousness began to follow a path similar to that of black power in objecting to pejorative media images and seeking land rights. A few indigenous people adopted the alternative culture that at times adopted them. McLaughlin and Davidson describe the Bear Tribe Medicine Society: 'Ten adults striving to relearn proper relationship with the Earth Mother, the Great Spirit, and all living things.... Share earth awareness and country-living skills ... Medicine Wheel Gatherings' (McLaughlin and Davidson, 1985: 351); of the ten, Sun Bear, who founded the group in 1971, is a Native American (McLaughlin and Davidson, 1985: 267–72).

In a different context, Alain Martineau writes of a transition from literary utopianism to a sense of how else things might really be:

> During the eighteenth century, utopia continued to refer to the limited meaning of an imaginary island ... In a larger sense, however, it expanded to include a universal vision of social life, radically opposed to existing social reality and its system of values. Utopia came to represent an awareness of rupture between social reality as it ought to be, and as it was. And, finally, it became the rejection of social reform to the extent that it represented a desire to begin *ex nihilo*.
>
> (Martineau, 1986: 33)

But there is no beginning out of nothing. I cite Venn again:

> The theme of an emancipation and of a liberation returns [in Fanon]. It requires rephrasing in a new secular idiom. This new discourse of the desired future ... is inevitably inflected by the history of modernity, and the language that had developed with modernity to speak of emancipation and enlightenment. But it cannot be written except in the course of a rewriting of the history of modernity and the philosophical discourse which has nourished it, that is to say, in the course of a critical engagement with how the experience of modernity has been temporalized and monumentalized, and what it has signified.
>
> (Venn, 2000: 105)

6 Ecotopias
Frameworks

Introduction

In Chapter Five I examined the moment of a potentially utopian culture in 1967, drawing attention to aspects of what may have been a radically new social architecture. I noted a departure from the city to intentional communities, and the tribalism of the counter-culture. In this chapter and the next, I bring the argument forward to the 1990s and the rise of environmental issues in the agendas of both mainstream and alternative social formations. A number of questions arise in this and Chapter Seven (in which I investigate anti-roads activism and ecovillage living), among them whether social justice is a precondition for environmental justice, or whether the latter can be considered independently of the former. This echoes the question as to whether social transformation is produced in shifts of individual or mass-social consciousness. A third question is how the experiences of grassroots environmentalism in the non-affluent world (called the global South) may inform environmentalism in the affluent world (called the West).

In this chapter I focus on critical frameworks articulating and arising from ecological and environmental debate up to the 1990s, including perspectives from the non-affluent world, dealing with issues of social and environmental justice that contribute to what might be called an Ecotopia, or green Utopia. I begin by considering grassroots environmentalism in the non-affluent world, ask how different concepts of the environment relate to different attitudes to the natural and social worlds, and move to the question of what kinds of knowledge emerge within green social formations. I structure the latter according to categories of principled (consciousness changing) and purposive (policy changing) intentions, and inward- and outward-looking formations, deriving these categories from Bronislaw Szerszynski and Andrew Jamison. This is not to say green knowledges should conform to such categories, which overlap and blur into each other, but it is a convenient way to open what I think may be an important question as to whether the shift of consciousness that occurs or is desired in environmental activism is personal or public – reformulating the aporia of individual and social change noted above. I take this up again in Chapter Seven. This chapter can be read, too, as background material for Chapters Eight and Nine on the alternative Modernism of mud-brick architecture, and barefoot environmentalism.

Grassroots resistance

Environmentalism did not begin only in the counter-culture of the 1960s. It assumes a particular form in land- and water-rights campaigns, protests against deforestation, and efforts for empowerment by groups in the non-affluent world, which parallel the civil rights campaigns that took place in North America in the 1960s, and movements for national liberation and cultural identity in a post-colonial context. Examples include resistance to a series of large dams in the Narmada Valley, India (Roy, 2001); blockades of forest clearance in Sarawak (Idris and Peng, 1990); the work of Bolivian peasant unions to halt soil erosion (Zimmerer, 1996); protest at cash-crop timber extraction in Karnataka, India (Guha and Martinez-Alier, 1997: 6–11); new farming methods developed by women's and other local groups in sub-Saharan Africa (Pradervand, 1990: 146–52); and efforts of farmers and activists to protect indigenous seed supplies and sources from the global monopolies of GM seed (Shiva, 2000; Yapa, 1996). Pierre Pradervand reports the following from meeting a farmer in Burkino-Faso:

> Halidou first spoke to me with quiet compassion of the farmer's self-image . . . He then explained how, despite the opinion of professional agronomists who told him that because of the region's hot climate it was impossible to store potatoes . . . he had invented a unique storage cellar that included an original ventilation system.
> (Pradervand, 1990: 149)

Ramachandra Guha and Juan Martinez-Alier write of opposition in Karnataka:

> Men, women and children took an oath of non-violence in a school yard, and then proceeded for a novel protest, termed the Kithiko-Hachiko (Pluck-and-Plant) *satyagraha*. Led by drummers, waving banners and shouting slogans, the protestors moved on to the disputed area. Here they first uprooted 100 saplings of Eucalyptus before planting in their place tree species useful locally for fruit and fodder. Before dispersing, the villagers took a pledge to water and tend the saplings they had planted.
> (Guha and Martinez-Alier, 1997: 9)

Guha and Martinez-Alier list four grassroots tactics: a showing of mobilization, usually in cities; disruption of the economy in strikes and blockades; direct action at specific sites or aimed at specific actors in the system; and finally hunger strike if necessary. They add, 'Mahatma Gandhi provides the environmental movement with both a vocabulary of protest and an ideological critique of development in independent India' (Guha and Martinez-Alier, 1997: 15).

Bina Agarwal observes that women have played a significant role in environmental protest. Asking whether villagers or women should be seen as victims of environmental degradation, she responds:

The emergence of grassroots ecology movements across the subcontinent (and especially in India) suggests otherwise. These movements indicate that although poor peasant and tribal communities in general, and women among them in particular, are being severely affected by environmental degradation and appropriation, they are today also critical agents of change. Further, embodied in their traditional interaction with the environment are practices and perspectives which can prove important for defining alternatives.

(Agarwal, 1998: 208–9)

Agarwal cites women's action in resistance to deforestation and construction schemes that damage village environments in Bihar, Kerala, Andra Pradesh and Garhwal, and argues that a bias against the poor means that deforestation and appropriation of village commons such as groundwater leave a majority dependent on shrinking natural resource bases while community resource management is eroded, resource degradation produces educational decline and high infant mortality, and population growth increases pressure on natural resources (Agarwal, 1998: 201). All this suggests that social and environmental justice are inseparable but that they mean different things to different groups.

To examine this further I look at the Chipko movement, an Indian campaign against forest exploitation which began in 1973 when villagers in the province of Garhwal demonstrated against the felling of ash trees for a foreign sports goods company. As the fellers approached, the villagers committed themselves to hugging trees (though in the event they did not have to), using the non-violent, bodily intervention employed by the Gandhian movement prior to independence. Sunderlal Bahuguna sees village disempowerment as beginning in colonial times when the British reserved forests for external commercial use. He cites a protest at Yamuna in 1930 when, after Gandhi's salt marches, villagers established a parallel government in Tehri-Garhwal, and were fired on by British troops who killed 17 of them. He adds, 'After independence, this spot became the place of pilgrimage for the hill people. They used to meet there to pay homage to the martyrs ... In 1968, they declared this day [30 May] as forest day and approved a charter of forest rights' (Bahuguna, 1990: 113). In time, the campaign against deforestation spread to other regions of India, and informed campaigns in the West (Bahuguna, 1990: 115; see also Guha, 1989). It became a movement for social re-formation as women adopted a specific stance within it. Agarwal notes that women in Chipko acted as watchwomen, selected species for reforestation and at times acted against men who prioritized cash income from development (Agarwal, 1998: 209–10). She adds that while Chipko draws on Gandhian traditions, 'women's responses go beyond the framework of that tradition and come close to feminist environmentalism' (Agarwal, 1998: 211).

Haripriya Rangan writes that once Chipko had gained media attention and achieved some government support, recurrent patterns appeared among its supporters. Village leaders and student activists affiliated with the Indian Communist Party, for instance, demanded higher wages for forest workers while Gandhians sought subsidies for local craft industries (Rangan, 1996: 215). Rangan draws attention to the role of a local elite (including Bahuguna, a forestry contractor),

94 *Practices*

arguing that at a crucial point Bahuguna framed the case for forest preservation in terms of a pre-existing national agenda of border defences and protection of national assets, rather than of local empowerment. She continues: 'politicians praised Chipko as the moral conscience of the nation' while Bahuguna gained international media attention, speaking from an 'interstitial space created by the state institutions, markets, and civil society' about 'simple, peasant ways of life' (Rangan, 1996: 216). For Rangan this glosses over 'the heterogeneity of classes, interests, and constituencies within the movement' (ibid.). If Chipko became an icon of grassroots resistance, its narrative retold around the world (Bahuguna accepting a Right-Livelihood Award in 1987), it was at the cost, for Rangan, of casting hill people as 'victims of natural disasters, the state, and the market' (ibid.). The campaign against deforestation has since expanded to a call for a new state of Uttaranchal, but Chipko's success in restricting forest exploitation has produced a decline in the need for labour and a halt to other kinds of development such as road building and irrigation, while increasing the power of central and state government. Rangan concludes:

> Environmentalists from India and elsewhere, rapt and slavish in their adoration and assiduous pursuit of romance with Chipko's ecological reincarnation, have been oblivious to the processes of marginalization continuing in the region. ... Paeans of praise for Chipko have drowned out militant local calls for tree felling rather than tree-hugging.
>
> (Rangan, 1996: 222)

It seems, then, that not only are the issues complex, affecting different groups in different ways, but that the success of grassroots resistance dependent on Western media attention or the involvement of local or Western elites distorts agendas. Cases of deforestation also highlight the intricate interaction of social and environmental agendas as local needs for materials, employment and autonomy meet globally perceived needs for preservation of vulnerable landscapes. Between external exploitation or calls for preservation, and local use and calls for empowerment, are what have been called endogenous development and intermediate technologies that can be developed and managed at grassroots level by local activists working with non-governmental agencies which are willing to hand over control of agendas as well as resources to local groups.

Liberation ecologies

The above quotation from Rangan raises difficult questions about development when it is seen from a non-Western viewpoint and does not mirror Western attitudes to the land as a site of wilderness or preservation. Further questions include whether development exhibits a bias against the poor, as it moulds the non-affluent world in the image of the affluent world, and whether some forms of aid prolong the situation they are designed to resolve. Jennifer Elliott remarks that after decades of development, developing countries 'have debt burdens which outweigh their Gross National Product several times over' (Elliott, 1999: 12). A division between the affluent and non-affluent worlds is no more precise, however, that that between social and environmental justice, inasmuch as the affluent world contains pockets of abjection just as the non-affluent world has its own elites with the same degree of mobility, consumer power and influence as other elites in the affluent world. Issues of race and gender arise in affluent and non-affluent contexts, too. Kate Berry argues that racial bias influences the access to natural resources such as water of indigenous Americans in their dealings with state institutions and companies (Berry, 1998, citing Levine, 1990; Meyer, 1984), which suggests affinities between the campaigns of non-affluent groups in the affluent world and groups in the non-affluent world. But a dimension specific to North America is that environmental issues raised by black groups in the 1960s were classed as social, not environmental issues. For Stephen Sandweis, environmental movements were 'overwhelmingly composed of white, middle- and upper-class members' whose key focus was the preservation of natural landscapes and endangered species (Sandweis, 1998: 38). Later, civil rights campaigns redirected green campaigning towards the idea that pollution is a denial of human rights, adopting a concept of environmental justice echoing that of social justice. In this scenario, and given the extent to which local campaigns are globally networked, it may not be the case that the local is inevitably overpowered by the global. But, in another way, the image of the Earth as a global entity, literally a globe, is problematic.

Images of the Earth seen from moon exploration vehicles produced a new idea of the planet as a single, unified entity. This might be expected to produce holistic thinking about the environment of what appeared a rather beautiful, unique globe of blue and green, spinning in space. But this nice image does not assist the localized campaigns of activists in poor countries for whom the planet is not whole but divided into the property of those with wealth, and the habitat of those who lack it (or whose inheritance is stolen from them in economic colonialism). The whole-Earth image may, then, be a version of the liberal humanist myth of a unified self projected onto a view of the planet from outer space. Arturo Escobar argues that when these images first appeared in the 1980s the dominant view of the environment moved from conservation to planet management. To the view that peoples and nations should undergo equivalents of the West's agricultural and industrial revolutions – posed in the September 1989 issue of *Scientific American* in answer to the question 'What kind of planet do we want?' (Clark, 1989: 48) – Escobar retorts, 'the question in this discourse is what new manipulations can we invent to make the most of nature . . . But who is this "we" who knows what is best . . . the (white male) Western scientist-turned-manager' (Escobar, 1996: 50). Raff Carmen similarly reads development as a myth, or modern 'surrogate religion' (Carmen, 1996: 11, citing Traore, 1990: 29), opening a question as to how iconic images, or narratives based on them, contribute to environmental awareness or conformity to established norms. Arren Gare sees a lack of 'stories of sufficient power and complexity to orient people for effective action to overcome environmental problems' (Gare, 1995: 140) and Escobar argues for 'new narratives of life and culture' (Escobar, 1996: 65). The question then is who will write the new narratives of place and social formation. Drawing on Manfred Max-Neef (1982), Carmen proposes that the means of environmental care and development should be handed over to the non-affluent societies which are conventionally the objects rather than subjects of development, which echoes the work of Paolo Freire (1972) in facilitating members of adult literacy classes to tell their own stories, and Augusto Boal (1979) in encouraging members of theatre audiences to walk onto the stage and take over parts at will (see Chapter Nine). Such an environmentalism of empowerment requires appropriate means, and attention to the role of local elites raised by Rangan (above). It rests on three grounds. The first is the efficacy of local knowledges (Guha and Martinez-Alier, 1997; Pradervand, 1990), and women's knowledges (Carney, 1996; Kabeer, 1994), and a capacity for mobilization produced when local groups work with 'social activists with the experience and education to negotiate the politics of protest' rather than alone (Guha and Martinez-Alier, 1997: 12). The second is the translation of stories of the negative impact of development into narratives of cultural identity that avoid a victim mentality. The third ground is the awareness that class, race and gender affect environmental issues. John Bretting and Diane-Michelle Prindeville find 'a relationship between social class, race, and gender and the exposure of individuals to environmental problems', to which responses include creating access to information, shared decision making, and being able to determine agendas, so 'community members exercise self-determination and learn advocacy skills' (Bretting and Prindeville, 1998: 141, 149). This is the radical environmentalism of the World Social Forum, calling at its 2001 meeting in Porto Allegre, Brazil for 'a trading system which guarantees full employment, food security, fair terms of trade and local prosperity' (Houtart and Polet, 2001: 123). It is echoed in direct action in trashing GM crops and disrupting fast food outlets in the West (Bové and Dufour, 2001). In the discourses of development and political economy, it is a basis for liberation ecology as a category of political ecology (Peet and Watts, 1996; Keil et al., 1998). Michael Watts and Richard Peet define liberation ecology thus:

> Liberation ecology integrates critical approaches to political economy with notions derived from poststructural philosophy. The quest is to understand the ways human practice transforms the Earth and the ways in which environmental practices, institutions, and knowledges might be subverted, contested, and reformed. [It speaks to . . . a critical analysis of environmental degradation and rehabilitation framed by something called development, and also the liberatory potential of struggles and conflicts exactly around these processes. Liberation ecology starts from Marx's presumption that society-nature relations are the outcomes of metabolic activity of the labour process . . . but posits this metabolism as social, cultural, and discursive . . . linked to the complex ways in which power, knowledge and institutions sustain particular regimes of accumulation.
>
> (Peet and Watts, 1996: 260)

Among its possibilities are a revision of instrumental rationality as the mechanism of exploitation of people and land; a questioning of the dominant voice in environmental policies and practices; and introduction of appropriate ways to hand over the position of determinant to groups and networks engaged in refiguring their social and environmental futures. This may produce the narratives that Escobar, Carmen and Gare read as the building blocks of a new ecological consciousness. A key concern is that empowerment – which is not donated by those in power – enables the use of appropriate knowledges and creation of an appropriate imaginary for plural, culturally and socially sustainable future visions. For Peet and Watts, as for Theodor Adorno and Max Horkheimer (1997), it is not a matter of junking rationality, but revising it from within:

> A liberatory ecology thus retains the modernist notion of reasoned actions, accepts much of the Marxian and poststructural critique of capitalist rationality, yet wishes to substitute for it a democratic process of reasoning – a sort of environmental public sphere – in a transformed system of social and natural relations.
>
> (Peet and Watts, 1996: 261)

I agree. But what would an environmental public sphere be like? How would it work? I take up these questions under the heading of 'Green knowledges' on p. 107.

Greening everyday life

Liberation ecology is a product of discourse between critical analysis of environmental campaigns and experiences narrated from the non-affluent world or non-affluent positions in the West. A shift has taken place, too, within Western institutions and professions since the 1960s, in advocacy and action planning, for instance, for an identification with local or marginalized publics. Radical planners (Sandercock, 1998) and development architects (Hamdi, 1995) are informed by the experiences of non-affluent publics in both the affluent and non-affluent worlds. Robert Chambers describes participatory rural appraisal in non-governmental development work as combining a sharing of information in which the knowledge of local groups is valued with a new 'willingness by professionals to be self-critical', so that as local people share knowledge with outsiders they learn more about their own knowledge, and thereby 'present and build up more than any one alone knew' (Chambers, 1998: 139). He cautions against empowering members of local elites, adding that the knowledge produced in dialogic encounters belongs to participants not external experts. A challenge for professionals in the West is to map the lessons of such work onto the processes of urban development at home in a socialization of the determination of urban futures. Parallel to that might be a shift on the parts of affluent-world publics participating in and assuming responsibility for environmental sustainability. As mentioned above, one area in which the latter takes place is green consumption.

The idea of a green lifestyle has begun to take root among consumers, with the use of green energy, recycling, second-hand or local economies, eating locally grown or organic food, and in some places voting for a green party. Green consumption tends to be seen as middle class but even if subsumed in a culture of consumer choice it has a positive impact on producers. Few people are able to move to an ecovillage or be green activists; more can buy organic vegetable boxes. For the most part, the environmentalism of the affluent world operates incrementally rather than through revolt, and at an individual or localized rather than a national level. This may mask the roles of industries and governments, and the decline in regulation in a period of globalization (Bauman, 1998), but change in the consumption behaviour of a mainly middle-class public has begun to shift supermarkets and companies towards adoption of ethical policies, or stocking fair trade and organic products. For consumers, buying organic carrots and potatoes may be to buy into a value system they find more beneficial than that of agribusiness. Jörg Dürrschmidt observes that people participating in organic vegetable box schemes do so to be part of a network, and as a rediscovery of local cultures through fear of globalization. This is 'a symbol of an alternative lifestyle, rather than an element of an actual lifestyle itself' (Dürrschmidt, 1999: 143). Contributions to local recycling, reductions in home energy use and using public transport are also symbolic (as well as practical) acts, dwarfed by the operations of transnational companies in a global economy, but these are the policy areas in which local and national governments retain influence.

I refer to this sense of belonging to a movement, without actually joining one, under the categorization of green knowledges on p. 107 below, and here would add only that allied to moves to sustainable architecture (described in Chapter Four), and largely outside the realm of a conventional politics of representation, green and fair trade consumerism appear elements in an incremental change towards sustainability, and can be compared to the long revolution through cultural and educational institutions proposed by Raymond Williams (1961). Williams saw the long revolution, which is of unspecified length, not as an alternative to political activity but in a relation to it, so that as incremental shifts of attitude are produced in culture and education, for example through the growth of adult education (to which Williams was a key contributor in Britain in the post-war period), so political attitudes would undergo a re-formation as well. The relation is mutual, in that political shifts also enable cultural and educational change; but the long revolution proceeds quietly despite the localized and ephemeral changes of conventional political (or party) structures, affecting these as well. Looking back over the six decades of the post-war era, the long revolution has produced possibly profound changes in social attitudes of race, gender and sexual orientation. Perhaps single-issue campaigning is part of another kind of long revolution, in which conventional politics (but not the issues of social justice adopted by the old political left and successive phases of a New Left) will be outmoded in favour of more direct means of political expression. This shift within mainstream society, in one way through green consumption, is complemented by activism and ecovillage living (see Chapter Seven), options not perceived as open to a majority though, as I argue below, in their way also a transformation of everyday lives.

Counter-affluence

Outside a politics of representation that is always symbolic are environmentalist and anti-capitalist activism Both began in single-issue campaigning in the 1980s on animal rights, the closure of United States air bases in Britain, and anti-roads protest. I read the emergence of these campaigns as a counter-affluence like the tacit revolt of the counter-culture of the 1960s (described in Chapter Five). Single-issue politics lacks the multi-issue programme presented by Leftist or Rightist parties; but party programmes lack the depth of analysis of single-issue campaigns, do not attract the same coalitions of interest and are bound to vested interests. Between the two is the imminent revolution in which localized shifts of attitude contribute to a long-term, cumulative shift of social formation to which I refer above. Histories of revolution make encouraging reading for those in power, however, as do histories of avant-gardes (Miles, 2004a), while the culture of personal liberation of the Summer of Love did not produce a new society but was dissipated in the atomism of the 1970s. The question was whether to change the world or one's consciousness. It recurs. Susan George writes:

> If neither mass personal transformation nor one-off revolutionary change can be counted on to create another world, perhaps we need to be a bit more modest. If capitalism does one day suffer defeat, I believe it will be as the cumulative result of hundreds of struggles, not of some great global apocalypse ... we have to keep on working to change the balance of forces in the most imaginative and non-violent ways.
>
> (George, 2004: 96–7)

But John Zerzan writes that nature as constructed in mass culture is 'a reminder of how deformed, non-sensual, and fraudulent is contemporary existence', breeding conformity in voting, recycling, and a pretence of normality (Zerzan, 2002: 159). He adds:

> Civilization, technology, and a divided social order are the components of an indissoluble whole, a death-trip that is fundamentally hostile to qualitative difference. Our answer must be qualitative, not the quantitative [solutions] ... that actually reinforce what we must end.
>
> (Zerzan, 2002: 160)

David Pepper notes that greens argue that Ecotopia should be a 'culturally diverse place where each small community lives out its own values'. He himself is sceptical, however, that the rationality of environmentalism – reaction to the 'irrationality of social injustice and environmental degradation' that do not produce benefits for most people – produces change; he believes in 'humanist, egalitarian and socialist aspirations' as prerequisites if Ecotopia is to be other than 'a repressive dystopia' (Pepper, 1996: 323–4). This can be borne in mind in relation to deep ecology as a potential eco-fundamentalism. A difficulty in understanding how Ecotopia might be produced is that the terrain is as divided as the quotations above show, while groups calling for social and environmental justice do so in a relation to dominant socio-economic structures like that of sub-cultures to mainstream culture. They are oppositional, seek a vocabulary that symbolizes their departure from the mainstream, yet inevitably carry within their opposition the terms of the departure set within the dominant society. There may be no exit from this impasse. I want next to look at two areas of the 1990s equivalent of a counter-culture, in the World Social Forum, and deep ecology.

The World Social Forum

François Houtart writes that the ideal of contemporary radical campaigning is the end of capitalism – the economic system that makes the global rich richer and the global poor poorer. He is clear that what is sought is an alternative to capitalism, not a revision of one of its historical forms. The demise of what he calls the 'real socialism' of the 1917 Revolution in Russia poses specific questions, among them the need for alternative movements to avoid suffocation in bureaucratic or ideological institutionalism and a need to re-examine power relations, but, as he also remarks, 'We should not forget that it is the very existence of the Eastern bloc, with all its ambiguities, which indicted Western societies . . . to establish the post-war welfare systems' (Houtart, 2001: 48). When neo-liberalism ignores or denies social relations, their recovery is itself a tactic for an alternative social structure. This is where experiences and histories from both the non-affluent and affluent worlds merge, in the analysis of class, race, and gender relations, but also in analysis of the 'precapitalist relationships between castes, different ethnic backgrounds, men and women' that underpin conflicts in the South (Houtart, 2001: 49). Houtart sketches the utopian alternative in this global context:

> We can dream of a perfectly balanced society, where the differences between individual initiative and solidarity are reduced to a simple state of tension, where human beings are judged because of what they are rather than the added value they produce, where cultures are considered to be equally valid expressions of being, and where scientific and technical progress is directed towards the well-being of all rather than the enrichment of a few.
>
> We *must* dream of this type of society . . . because even if it is not attainable in our *topos* (place), it does have the force of attraction, mobilizing the spirit and the heart in this dream of the necessary utopia. But unless this utopia starts out from a firm conviction that it is *possible* to construct another social logic, and thus to approach the ideal, it remains a dream.
>
> (Houtart, 2001: 51)

Utopia, then, is a an image to grasp, realized in a faith that it is really possible (the real-possibility of Marxism). If sustainability is 'a quasi-magic term' (Houtart, 2001: 53), it excludes social relations, and the term eco-development is therefore coined to denote a creation of relationships of social production in response to environmental degradation and economic deregulation, achieved through grassroots action and coalition building. Further, Samir Amin argues that a new pluralism arises when local struggles are globally networked to link landless farm labourers in Brazil, workers and unemployed people in Europe, salaried staff in medium-developed countries such as Korea and South Africa, and student activists, all in global and cross-class coalitions. Recognizing the impact of new social movements, Amin proposes a fusion of efforts towards social democratization and economic determination as 'the main strategy around which struggles can unite' (Amin, 2001: 61). In 2001 at Porto Allegre, a Call for Mobilization in face of globalized capital states that neoliberal globalization:

> destroys the environment, health and people's living environment. Air, water, land and peoples have become commodities. Life and health must be recognized as fundamental rights which must not be subordinated to economic policies.
>
> (Houtart and Polet, 2001: 123)

At this point environmentalism and socialism fuse in an emerging praxis, the global equivalent of single-issue campaigning, and a new politics (or re-politicization of the issues hitherto displaced to the private operations of a free market). Meanwhile, groups of anarchists and socialists continue to organize and theorize after the demise of state socialism (Albert, 2001; Barker, 2001); and a new political form called dis/organization (Jordan, 2002: 53–80; see also Chapter Seven) appeared in the Carnival against Capitalism in London on 18 June 1999.

Deep ecology

An interiorized response to global environmental degradation is the rise of deep ecology as a strand in Western philosophical and cultural discourses (Sessions, 1995; Naess, 1989). Deep ecology replaces anthropocentrism – the centrality of the human in the human world view – by ecocentrism – a centrality of ecologies as interdependent living systems – as the paradigm for human development. Fritjof Capra write of a holistic world view that requires a deeper ecologism than a greening of mainstream society – shallow ecology in remaining anthropocentric – and 'a corresponding ecologically oriented ethics' (Capra, 1995: 20). This, he asserts, replaces the mechanistic view of the seventeenth century, caricatured as representing the universe as a clock, by a model in which natural systems arise from interactions and interdependencies that are specific and not known by dissection, while 'the nature of the whole is always different from the mere sum of the parts' (Capra, 1995: 24). Capra charts a corresponding shift of values from expansion, quantity, competition, and domination to conservation, quality, cooperation, nonviolence'. Norwegian philosopher and mountain-climber Arne Naess proposes an eight-point plan (Naess, 1995: 68), which I paraphrase:

- life (both non-human and human) should be valued in itself;
- biodiversity contributes to that value, and is also a value in itself;

- humans have no right to reduce diversity except in vital need;
- the success of the human species is compatible with a flourishing of non-human life;
- current human intervention is excessive;
- radical economic, technological and ideological policies are required;
- a lifestyle of inherent value should replace that of growth;
- those who subscribe to the above should act accordingly.

Naess defines the differences between deep and shallow ecology in part in the approach to population density, which 'must have the highest priority' in the industrialized world as well as elsewhere (Naess, 1995: 73). He does not say how it is to be addressed, or by what voluntary or persuasive means the populations of other countries are to be reduced, and commentators from the non-affluent world such as Rangan argue in any case against the thesis that poverty follows over-population rather than being its cause. Naess also advocates preservation of pre-industrial economies, limits on Western technological exports, and a new kind of education to promote the idea of a (non-quantitative) quality of life and reduce consumption. For Naess, this leads to a process of personal self-realization, explained in complicated diagrams and a terminology of his own that, obliquely and coincidentally, might be compared to the intricacy (but not the content) of Charles Fourier's utopian imagination (see Chapter Three).

Deep ecology can be understood in the context of an appreciation of the remote, white landscapes of Norway – a rich state with a small population – but has gained attention in the ecovillage movement (Bang, 2005: 50). The idea of being lost in a landscape has a cultural history, not least in the exploration of the (as if unpeopled) west in North America and the project of voluntary simplicity developed by Henry David Thoreau and described philosophically in his account of living alone in a log cabin at Walden Pond (Francis, 1997: 218–49). Richard Francis observes:

> Thoreau envisages a day when the absolute laws of nature will permeate and finally replace the contingent reality of daily life . . . when history and nature will be reconciled and become what he calls mythology. Because he is thinking in individual rather than communal terms, however, he finds it very difficult to see how this occurrence might be part of social development.
>
> (Francis, 1997: 229)

Francis indicates Thoreau's response earlier in his text, noting his attitude to architecture as 'decrying ornamentation and the creation of external surfaces that fail to reflect the inner core' (Francis, 1997: 224). Thoreau asserts that among the most natural of human acts is the building of a house, an act that dissolves the division of labour when a single individual possesses the skills necessary for life – cultivation, shelter, clothing, exchange and spiritual nourishment – as personified by the farmer, builder, tailor, merchant and preacher in everyone (Francis, 1997: 224). Thoreau thereby regards himself as society in microcosm, neatly dissolving the problem of the individual and society.

I find this problematic, not only romanticized but contradictory in that, despite the appeal to a natural state to which the solitary dweller aspires (and sees mirrored in wilderness landscape), the wilderness is exactly that, a landscape framed by a specific human culture at a specific time. Similarly, regardless of its protestations, or a particular kind of Protestant frugality perhaps, deep ecology's ecocentrism is another form of the human ordering of apperceptions of the environment. This, of course, in no way prevents deep ecology or Thoreau's solitude from being revealing objects of cultural criticism. In that context, Thoreau, as an iconic figure in a literature of solitude, shares with deep ecology a century later a vision of a life prior to industrialization. That vision raised real issues in Thoreau's time, when railways cut through the hitherto uncharted west; but today, the pre-industrial tends to be displaced to the non-affluent world – a world on which deep ecology has little to say beyond co-opting that, too, to wilderness. Here, paradoxically, is the idea which deep ecology shares with Thoreau's time: that the land, and in nineteenth-century North America its native inhabitants, are co-opted to an Edenic vision or Rousseauesque abode of the noble and uncorrupted savage. To white settlers this savage is, of course, ripe for colonization. I am not sure whether being the object of romanticized projection is much better.

Like the retreat to garden suburbs in London in the 1890s (discussed in Chapter Four) the retreat to solitude in wilderness is, though more extreme, reactive and regressive. Perhaps, too, the legacy of Walden Pond, contextualized by a North American dream of the log cabin as denoting purity, meets what I might call deep ecology-lite in a vogue for frugality that has produced its own popular literature. For example, Jim Merkel writes in *Radical Simplicity* (2003) that an interspecies, interhuman, intergenerational equity is required, as might be found in deep ecology, and frames his argument in appeals to indigenous wisdom and the methods of pop psychology. Citing Chief Seattle on 'the end of living and the beginning of survival when the scent of man [*sic*] began to permeate the fragrant wild lands of his people' (Merkel, 2003: 55), he offers a nine-step programme derived from the book *Your Money or Your Life* by Vicki Robin and Joe Dominguez (Merkel, 2003: 124). To introduce this material may seem frivolous, but I make the juxtaposition because Merkel and Naess privilege a withdrawal from society framed as a desire for wilderness that, in constructing the land as a landscape foil to human urban-industrial society, seems at root to appeal to a new version of the sublime, in the process evacuating the human subject of agency. This is not an argument against species equity, though that can be read in terms of an instrumental rationality whereby the survival of the human species depends on the preservation of complex ecologies, damage to any element of which threatens all others. Or it can be a matter of belief. Wanting to be rational but to find alternatives to instrumentalism, wondering how a vision of a depopulated planet can be a narrative conducive to a shift in human attitudes to caring for that planet, and from a feeling that much of what I have described above is romantic escapism, I need to look more carefully now into the varieties of a concept of the environment to understand how these shape the narratives and philosophies of environmentalism.

Environments and environmentalisms

The model of an ecological footprint by which individuals or states measure their use of non-renewable resources (Chambers *et al.*, 2000) offers a tool for the negotiation of environmental futures. A footprint implies a view of environment as habitat as well as finite resource, and is useful in indicating differences between levels of land use, energy and water consumption, and waste recycling between affluent and non-affluent societies. Instead of a focus on population numbers it offers an evaluation of the comparative efficiency of resource use: 'footprinting addresses not the number of heads but the size of the feet' (Chambers *et al.*, 2001: 59). Yet if footprinting advises more equitable use of the planet's finite resources in the interests of sustainability, this might be extended from a focus on production to a concern for new approaches to distribution. Frances Moore-Lappé writes in the context of Central America, 'nowhere more dramatically . . . can it be shown that hunger is caused by anti-democratic structures of power, not by scarcity' (Moore-Lappé, 1990: 131). There, footprinting will demonstrate inequalities as facts, when the rich use resources to excess, but not the lack of equitable structures of value as a political concern. When, however, the planet is seen as a mutually owned resource, in a way quite contrary to deep ecology's privileging of its spiritual or cultural image, this connects to ideas of mutuality in social organization, so that environmental justice might grow out of social justice. But while the environment is seen as a landscape, the pristine parts of which are to be preserved, then the question of ownership, and who decides whose land is to be preserved outside the cycles of use, are masked; at this point environmental preservation displaces environmental justice.

John Barry classifies the environment in four ways: as wilderness; as countryside or garden; as urban fabric; and as global system (Barry, 1999: 22–9). The four categories denote a historical trajectory from hunter-gather to farmer, city-dweller and traveller in a new world order. The most recent phase, the global, is used by Barry to indicate both a culture of risk and a view of the Earth from space (Barry, 1999: 27). Barry writes of the global economy as creating 'socio-economic relations between distant places and people to express 'the interdependence of distant people and places' (Barry, 1999: 27–8). This uncritically mirrors the view of the Earth cited by Escobar (see p. 95). Barry's tacit trajectory is problematic, too, in that the model of progress implied in a trajectory of socio-cultural development privileges the society regarding itself as the culmination, as Annie Coombes shows in a study of early ethnographic displays (Coombes, 1994). I would prefer to see Barry's four terms in a symmetry in which wilderness and the Earth seen from space are terms of sublimity, and the pastoral and urban are terms of what is cultivated. Part of what is cultivated is the idea of an environment: literally in that landscapes are produced in forestry, agriculture and national park conservation; and metaphorically in that they are framed by ways of seeing, which are political (Warnke, 1994). The preservation of wilderness itself is a political act requiring legal measures and enforcement, and reflecting the interests of particular elements within a society – hunters, hikers or climbers.

Different concepts of environment also produce different literatures. A growth in practical guides – indicatively, to alternative energy (Elliott, 2003), ecological house renovation (Harland, 1993), bio-regional planning (Desai and Riddlestone, 2002) and building in earth and straw-bale (Kennedy *et al.*, 2002) – suggest a growth in do-it-yourself environmentalism and a redefinition of needs differentiated from consumerist wants (see Papanek, 1984, 1995). Like green domesticity (Noorman and Uiterkamp, 1998), green revision of wants, as in green consumerism, provides demonstrations of possibility and lends participants a sense of belonging to alternative networks. A parallel tendency in non-governmental development work in the non-affluent world is the growth of intermediate technologies and local management of projects. This has been termed endogenous development: 'that is mainly, though not exclusively, based on locally available resources and the way people have organized themselves' (Haverkort *et al.*, 2003: 30). Issues emerge of means and ends, and the relation of appropriate and socially just means to different meanings of environment.

If the underpinning image of a natural environment is wilderness, the strategy may be depopulation. If it is the village common or urban allotment, it is more likely to be the negotiation of equity, at which point arguments for species equity may be reintroduced as an extension of social justice, or land rights – and even the rights of land and natural growth as well as non-human animate creatures – as an extension of civil rights. I wonder how much the visual plays a role in the way the environment is conceptualized. Escobar argues that the view from space 'was not so great a revolution. This vision only re-enacted the scientific gaze established in clinical medicine at the end of the eighteenth century' (Escobar, 1996: 49). I would add that a view from far away distances the object of perception in particular ways, as stated above, lending an illusory unity. Of course, all representation is symbolic; but it is exactly in reaction to (or regression from) this fact that a quest for authenticity identifies images of wilderness as a sign of a pre-corrupted Earth in which to find pure existence, appealing to bourgeois sensibilities as tribalism attracts people who drop out of mainstream society. There can be no going back, but what would a natural reclamation of the Earth be like? This is shown in Richard Jeffries' novel *After London* (1885 – see Jeffries, 1895). The weeds have taken over and the pavements have disappeared after an unstated catastrophe has emptied the city of people: 'By the thirteenth year there was not one single open place, the hills excepted, where a man could walk, unless he followed the tracks of wild creatures or cut himself a path' (Carey, 1999: 278). Jeffries' imagery is ambivalent: he describes a flourishing of wild plants but also the survival of the strongest species pushing through, in a post-Darwinian image of an overgrowth without design. His book also resembles late twentieth-century fictions of post-nuclear survival – such as a television programme called *Survivors* in which a band of post-nuclear refugees wandered Britain, encountering other groups in competition for scarce resources by which to re-establish society – and may have a symbolic appeal similar to that of courses in survival skills. A half-page magazine advert offers 'training courses in wilderness skills and bushcraft/survival'; the teacher has experience in scouting, military service, pest control, craft and woodland management. The advert (in *Permaculture*, 49, Autumn, 2006, p. 61) is illustrated by knives, axes, and possibly a crossbow in the lap of a man sitting by a camp fire. Suzi Gablik records a conversation with artists Rachel Dutton and Rob Olds in 1992 as they were about to undertake such a course. Olds' final words were: 'Hunter-gatherers are the apex of human civilization. We need to go back to that point. . . . we can't assume what is comfortable for society . . . You've got to do it all the way, or not at all. Or we die' (Gablik, 1995: 83). The trajectory of hunter-gatherer, farmer and urbanite is upended by Olds, but perhaps the difficulty is not the linear direction, back or forth, but the model of trajectory itself.

The model of a trajectory is used by German Idealism in the nineteenth century, in which the ultimate point is freedom as the realization of Reason, and informs the model of a rising (or uprising) working class in Marxism. It is incompatible, however, with the non-teleological model of Charles Darwin's *Origin of Species*, first published in 1859 and implicitly a context for Jeffries' novel. Elizabeth Grosz comments that Darwin says little about the origin of species and writes instead on the mutation of species, which is non-teleological. Darwin, she notes, takes the development of verbal languages as a similarly non-teleological, varied process based on three principles: 'individual variation, the reproductive proliferation of individuals and species, and natural selection' (Grosz, 2004: 32). The evolution of life forms and languages depends, then, on an irreversible temporality requiring 'abundant variation; mechanisms of indefinite, serial or recursive replication or reproduction'; and 'criteria for the selection of differential fitness' (ibid.).

This has two implications for a concept of a natural environment: that it is produced in a non-reversible sequence that has neither design nor end; and that it is complex, as in an ecological system of numerous elements linked by innumerable relations, in which a change in one element leads to changes in the others. From complexity theory (Cilliers, 1998) it follows that the outcomes of change cannot accurately be predicted, for a slight change in the conditions of an intervention produces major changes in its outcome. The Darwinian model denotes a transformability of elements in the planet's ecosystem, and the impossibility of stasis – 'the impulse towards a future that is unknown in and uncontained by the present and its history' (Grosz, 2004: 32; see also Pepper, 1996: 258–9). This invalidates claims to return to a natural state or reassertion of naturally given rights. The state of origin is irrecoverable, and unknowable. If the biological future extends from but is uncontained in its past, the same may be said of human history and verbal language, at which point the dualism of nature (biology as non-historical) and culture is collapsed. If biology is produced, it is cultural. If biological culture is non-teleological, so by allusion is human history (though allusion is a speculative mode of argument). A difficulty arises, then, in a trajectory from wilderness to the whole-Earth image, in that wilderness is not an innocent or original ground, and the whole-Earth image, which is obviously a cultural construct, is not a return to a unified, holistic Earth that once existed and was corrupted by human intervention. Like the noble savage, that ur-state never was. Is Ecotopia to be made new, then?

Green knowledges

Finally, I ask what knowledges are produced by different environmental constructs. Andrew Jamison uses Bronislaw Szerszynski's categorization of green attitudes in four kinds of cognitive praxis (Jamison, 2001: 149–50). Szerszynski calls these four pieties, which I paraphrase:

- a monastic piety of principled, counter-cultural groups who develop green lifestyles, withdrawing from the dominant society;
- a sectarian piety of purposive counter-cultural activity, like the Puritans of the seventeenth century witnessing against evil;
- a churchly piety that is purposive and established, occurring in organizations and environmental campaigning groups like Greenpeace;
- a folk piety that is principled and mainstream, as in green consumption.

I return to this in Chapter Seven, arguing that activism is principled (seeking to shift the consciousness of those taking part in the moment of action), not purposive as Szerszynski says. I find the four-part categorization helpful, though, in differentiating a desire to find liberation, personally and in groups, from a wish to rewrite political agendas.

Jamison adapts these categories to provide another set, of professional, personal, community and militant action. Professional and community environmentalists are purposive, focusing on outcomes, and are instrumentalists – they do one thing to make another thing happen. But militant and personal environmentalists are practitioners of alternative lifestyles in public and in domestic settings, respectively, who live in the new society. Jamison argues that militant groups are often splinter groups departing from institutional structures, infused with political ideologies in a tradition of protest that is 'characterized by their defence of "traditional" practices and ways of life' while their resistance is based on identification with 'small-scale farming or shopkeeping' (Jamison, 2001: 167). Jamison likens such groups to the residual cultural formations described by Williams in relation to rural societies (Williams, 1977). Mapping these categories onto resistance in the non-affluent world, militant groups protect and use local knowledges, seeds, and other kinds of material and immaterial culture. But many non-governmental development workers have, like radical planners, crossed over and started listening to or spending time with the grassroots community (see Sandercock, 1998: 97–104). This action values tacit knowledges as equal with intellectual and scientific knowledges, and is not a leap into mystical or irrational mentalities.

As Chaia Heller writes, in discussion of the socio-erotic:

> Assertions of irrationality or intuition as epistemologically more authentic or immediate than reason are predicated on the myth that reason is the opposite of intuition. However, intuition often constitutes a pre-reflexive expression of rationality: when intuitions are right, they reflect historically grounded insights

that we have rationally cultivated about the world; when they are wrong, they often reflect more about ourselves and our unconscious desires.

(Heller, 1999: 117)

Much of Heller's discussion might be compared to Fourier's idea of a libidinal society (see Chapter Three), except that Heller avoids the tabular presentation by which Fourier seems captivated. Heller emphasizes reinterpretation of events and environments as a means to empowerment, commenting that how news is interpreted affects how dysfunctionalities are addressed. Seeing the socio-erotic as relational and empathetic, Heller calls for 'non-hierarchical structures that facilitate *meaningful* cooperative social relationships in all areas of our lives' (Heller, 1999: 92). Among the desires that animate such socio-erotic relations is oppositional desire, in 'passionate and rational action to defy institutions that impede the creation of a just new world' (Heller, 1999: 109).

This brings to mind the sexual revolution of the 1960s, and the counter-culture, but Jamison writes that personal environmentalism has diversified since the hippies and the fledgling environmental movement of the 1960s, and that while 'members of the counterculture expressed themselves not just by wearing colourful clothes but also in their voluntary simplicity and a youthful openness to non-Western cultures that led many to move to the countryside and take journeys to the East' (Jamison, 2001: 169), a shift from New Age to another, more mainstream, personal environmentalism new occurs in the rejection of GM foods. This is linked to identity construction in a consumer society so that green politics are subsumed in identity politics. At the same time, environmental campaigning has become more organized, at times on a business model. In terms of pieties, the sectarian has departed further from the mainstream as the folk and churchly have grown within it. I would probably conflate the sectarian and monastic, as activism and ecovillage living, in the specifics of building ecological social architectures – as discussed in Chapter Seven.

I began this section of the chapter by asking what kinds of knowledge are produced when different concepts of environment are employed, and in different socio-political configurations. The question is not entirely precise, but implies that in specific contexts framed by the cultural concepts of a specific interest within a society, particular kinds of awareness are likely to emerge. The triangulation points by which those kinds of awareness can be described are Szerszynski's and Jamison's categories, as I redraw them, of principled-personal, principled-social and purposive. That is, the principled-personal is awareness of liberation in personal consciousness, as an end in itself and likely to be realized in a departure from the dominant society, though not necessarily in solitary contemplation – it could be in ecovillage living, perhaps. The principled-social is also a transformation in consciousness but in the moment of social re-formation – in activism, perhaps. There is no hard line between these two categories, and individuals and groups may move between their settings. The purposive conflates efforts to change policy with efforts to adopt alternative lifestyles within and retaining membership of mainstream society, as in supporting campaigning organizations and green consumption. In Szerszynski's terms this is a conflation of the churchly and folk pieties, which I justify in that both are, in their own terms, concerned with public policy, socio-economic and political organization, and are the means by which groups and individuals within a society incrementally shift its identity. The latter is the long revolution (Williams, 1961) in diverse formations; the former are more likely to be sudden, revelatory and immediate. This gives me two possibilities: for an imminent revolution through incremental shifts, within what is usually called a public sphere (in which members of a society collectively determine its values and their organizational enactment); and an immanent revolution that is closer to millenarianism but can occur in everyday life. I do not read these terms in opposition but as polarities between which there is a creative, discursive tension. In the next chapter I keep this in mind when considering activism and ecovillage living.

7 Ecotopias
Practices

In Chapter Six I set out a framework through which to consider environmentalism as it has developed since the 1960s. I drew attention to views from the non-affluent as well as affluent worlds, to issues of gender and race, and the variety of positions contained within the terms environment and environmentalism. Finally, asking what knowledges, or kinds of awareness, are produced in relation to different concepts of the environment, and perhaps more in different social formations that themselves produce different models of how the environment is approached, I proposed a transformative awareness, a shift of personal consciousness, and a public awareness articulating an incremental shift towards a greener future. I used the terms immanent and imminent revolution to denote these possibilities. I want to reiterate that these are overlapping, blurring categories introduced for the sake of argument in an effort to understand what meaningfully happens in the diverse ways in which an alternative society emerges. The model of imminence and immanence seems useful, and I keep it in mind in this chapter, while returning to the question as to whether activism and ecovillage living are arenas for the kind of awareness aimed at changing policies or the kind that constitutes a personal transformation.

In this chapter, then, I investigate the emergence of environmental activism and the growth of ecovillages in the 1990s. I note parallels between the ideas of recent environmentalism and the counter-culture and spread of intentional communities in the 1960s. For instance, the use of direct action by radical groups in the 1960s was extended in a new context by anti-roads protesters in the 1990s, and the tribalism of the Rainbow Gatherings has a parallel in 1990s anti-roads protest. There are also differences. If 1960s counter-culture appropriated aspects of indigenous cultures, 1990s activism was informed by local struggles against exploitation in the non-affluent world. And if ecovillages, like intentional communities, are departures from mainstream society, they nonetheless pilot low-impact ways of living from which mainstream society can learn. I draw out a small number of issues in activism and ecovillage living, and examine these by specific aspects and cases, from anti-roads protest in Britain and North America and the Donga tribe at Twyford Down to ecovillages in Europe and Australia. Among key concerns are the motivation, community and ecology of such settlements.

Activism

Perhaps activism offers a moment of an emergent new society. By a moment I mean, from Henri Lefebvre, a fracture in the dulling time of routine when a pervasive clarity or sudden realization occurs. I think this moment of a new awareness constitutes a new kind of knowledge produced in activism – a glimpse of a new society implicit in an awareness that takes place among others while experienced personally. This draws on Hannah Arendt's concept of natality – a mature self produced through perceptions among others (see Arendt, 1958, 1971: II, 109–10; Curtis, 1999: 93–123) – as well as Lefebvre (see Shields, 1999: 58–64). I propose it here speculatively (because I am not an activist) as a way to understand what happens in the acts of activist campaigns, and why this matters.

From this it follows that the ephemerality of activism is not cause for its dismissal, but a reason for its consideration as a new politics outside a politics of representation – an advent of direct democracy and spontaneous, informal organization. It is ephemeral in that an action lasts a day, but the new awareness produced in the consciousness of those participating is of lasting import. Similarly, activism emerges in waves and subsides, but continues to engage through the persistence of its networks. Derek Wall concludes, in his study of anti-roads protest:

> A certainty of social-movement theory is that waves of mobilization ebb and flow; they do not grow in a continuous and sustained fashion. Yet EF! (UK) [Earth First!] has promoted radical repertoires of action and forged a wave of green protest even after a period of decline in environmental activism. New targets for direct action have emerged. Direct action has become more acceptable to both greens and wider society, a move that must please most radical opponents of planetary destruction.
>
> (Wall, 1999: 192)

Tim Jordan argues that moments of pleasure generate an ephemeral politics, noting that punk was once resistant but became fashion. He continues that 'when such cultures are closely associated with an oppressed community and create representation for that community . . . the two – politics and pleasures – may be joined . . . collective pleasures become political by being connected to a politics and not by being a politics themselves' (Jordan, 2002: 81). This is evident in rave culture (Jordan, 2002: 81–101; Huq, 1999; McKay, 1996: 103–26), and localized campaigns in which people are drawn into new, oppositional collaboration.

Widening the context, cultural forms, especially the performative, figure strongly in 1990s activism, addressing repression through irony, wit and theatre. This was the case, too, when the San Francisco Diggers banalized racialism (see Chapter Five). Starhawk, a North American anti-globalization activist, writes of fluid street actions: 'claiming a space and redefining it, disrupting business as usual . . . embodying the joy of the revolution we are trying to make' (Starhawk, 2002: 151). Seeing a need for a positive image of humanity within a valuing of nature, she writes:

For if we believe we are in essence bad for nature, we are profoundly separated from the natural world . . . subtly relieved of responsibility for developing a healthy relationship with nature . . . The human-as-blight vision also is self-defeating in organizing around environmental issues.

(Starhawk, 2002: 161)

This implicit confidence in human goodness can be compared to Heller's model of the socio-erotic (1999) and Jordan's interest in the politics of pleasure (above). It coloured the Summer of Love and informs the free festivals in Britain of which George McKay writes in *Senseless Acts of Beauty* (1996).

Citing the Windsor Free Festivals of 1972, 1973 and 1974, McKay traces their roots to San Francisco, Woodstock and countercultural activity in Britain such as ripping down fences at commercial popular music festivals (McKay, 1996: 12–14). From an interview with their organizer, Bill Dwyer (a London squatter), he links the festivals to the action of the Diggers at St George's Hill (see Chapter Two). They follow, too, a growth of radical poetry in the culture of the Beat generation, and interest in William Blake as a subversive. McKay says of Mick Farren, organizer of the non-profit music event Phun City in 1970, that he discovered 'the utopian possibility of connecting counterculture with rock music' (McKay, 1996: 14). Citing Dwyer, he writes:

Tripping on acid in Windsor Great Park he [Dwyer] had a Blakean vision of a communitarian utopia, which he thought he could bring to life by holding 'a giant festival in the grandest park in the kingdom, seven miles long!' Why hold it at Windsor? Because it's an effort to reclaim land enclosed for hunting by royalty centuries before – an updating of the . . . Digger strategy.

(McKay, 1996: 16)

The Queen was invited but did not attend. A few thousand people arrived, camping in a copse. In 1973 there were tens of thousands, and the festival called for the legalization of cannabis. In 1974 Dwyer suggested that the United Nations should hold an ecological fair at the festival. The police intervened, and the third was the last Windsor Free Festival. By then music festivals were established – Glastonbury since 1971 – but were already losing their radical roots to become commercial events. The British counter-culture moved to Stonehenge and further confrontations with the police as its nomadism was viewed as a threat to social stability (McKay, 1996: 31–3). McKay quotes Elizabeth Nelson, author of *The British Counter-Culture 1966–73* (1989), that the young people who participated in free festivals did so as consumers, in imitation of the North American lifestyle of free concerts such as Woodstock. He objects, though: 'I don't think it's true of the British free festival, that illegal, subversive, and sometimes very dangerous place', and that Nelson misses the significance of 'the ancient British tradition of festivals, seasonal, nomadic, and otherwise' (McKay, 1996: 34). McKay argues that a 'possibility of a kind of change' was glimpsed in 1970s festivals and fairs which feeds into activism (McKay, 1996:

48). This was both future-looking and nostalgic. I recall, for instance, how much the name Albion was used in the 1960s for a mythical free Britain.

Perhaps there is a free-spirit millenarianism in activism and New Age travelling in the 1990s, too, though the term New Age is not recent but was the title of a journal produced by the Owenite community at Concordium in the 1840s. *New Age* promoted mysticism, the Love Spirit and vegetarianism (Pepper, 1996: 215). Wall cites influences from a related emergent counter-culture in the 1880s and 1890s on the British Labour Party, arguing that 'socialism in Britain was strongly linked to green and anarchistic sentiments', while green issues were articulated through pressure groups and rising public interest in nature in the 1890s and 1900s (Wall, 1999: 18). He adds, 'the first recorded environmental pressure group, the Commons Preservation Society ... combined disruptive direct action with legal challenges and political lobbying' (ibid., citing Marsh, 1982: 44).

Some New Age travellers in the 1970s were ex-squatters, taking to travelling as the process of eviction was made easier for property owners by the 1977 Criminal Law Act. Others took to the road as a practical response to economic need during recession, though as McKay points out, citing Jay, a traveller he interviewed, squatters were not affected by economic recession because they did not work and bought mainly second-hand goods (McKay, 1996: 47). There were also allusions to tribalism – for instance at Tipi Valley in Wales, established in 1976 by a group of the Hyde Park Diggers. Shannon, a visitor in 1980, recalls: 'It looked like some picture of the Rocky Mountains ... hundreds of years ago' (in McKay, 1996: 54). McKay reads its Native American appropriation as related to struggles for land rights in a 'self-conscious politicized choice' and re-politicization of dwelling space contributing to a 'romantic construction of alienated people' when '... the sheer otherness of this chosen living space' indicates rejection of mainstream society (McKay, 1996: 56) – as with the Farm (see Chapter Five). Shannon mentions internal tensions at Tipi Valley, and bags of rubbish littering the site. But nomadism, rejection of family ties, refusal of productivity, and community among others refusing the abjection offered by the dominant society, all fed into the anti-roads protest of the 1990s. In *Green Political Thought*, Andrew Dobson wrote in 1995 that anti-roads protest had become a prominent feature of the British political scene, linking 'an apparently disparate collection of people, ranging from middle-class "Nimbys" [not in my back yard] through to New Age travellers' (Dobson, 1995: 146). Dobson cites disillusionment with political parties and a desire to live self-reliantly, and quotes an unidentified activist that the isolation of do-it-yourself politics in squats and raves led to the emergence of 'tribes' who 'talk about a resurgence of the free-spirit movement' in a 'network of the skint but proud' (Dobson, 1995: 146). Patrick Field writes of unlikely allies in the first anti-roads success at Oxleas Wood, east London, 'outlaws nursing secret dreams of martyrdom and one "excuse-me-do-you-know-who-I-am" professional person, equipped with a mobile phone whose memory contains the numbers of several national newspapers and a lawyer' (Field, 1999: 73). Mick Thompson, living in Claremont Road, east London, the site of a motorway extension scheme and anti-roads protest in 1992–3, argues that constitutional means to object to new road schemes broke down in that period simply because the Department of Transport's programme was too widespread to allow the detailed and bureaucratic processes involved to operate effectively, while public inquiries were seen as paying no more than lip-service to objectors. He adds, 'Most of the campaigns were rooted in local communities, linked by umbrella groups . . . using background research and support from well-established environmental organizations' (Thompson, 1997: 37).

The eviction of the residents of Claremont Road took more than a year and gained national media attention. Artists renting short-life houses as live-in studios in the street, activists who squatted empty houses prior to the road scheme and a smaller number of long-term residents refusing to move, made Claremont Road a colourful and oppositional site. Sandy McCreery, an academic, writes that in 1994, during the period of attempted eviction:

> Claremont Road was transformed into an extraordinary festival of resistance. The houses were pulled apart and remodelled with the original components and anything else that could be put to use. One became the 'Art House' where visitors were invited to participate in, or view, a constant outpouring of murals, installations . . . Another house, converted into the 'Seventh Heaven Jazz Café', was particularly intended to attract day-trippers to the street . . . The exteriors of almost all the others were brightly painted with various images: a floral frieze . . . portraits of people in the street; dreamlike celestial horses; and political slogans.
>
> (McCreery, 2001: 231)

Rope netting was strung between trees, barricades were built, and a tower of scaffolding poles rose 60 feet above the houses like a public monument. Sundays were for parties. There was no formal organization but collective meals were eaten and work was done as needed on a voluntary basis. McCreery adds that 'the street was also a place of obvious good humour' (McCreery, 2001: 235).

There are several overlapping references in the above accounts. First, single-issue campaigns on animal rights and the protection of environments drew together young travellers with retired schoolteachers and people of diverse class backgrounds, but not always with the same tactics. Field (1999: 74) argues that the use of non-violence unified resistance while young protesters inspired older residents, but that occupations were like carnivals and the fortified camps like adventure playgrounds (an element of 1960s urban community arts). Residents of all ages and backgrounds supported or supplied protesters – the women's Peace Camp at Greenham Common, for example, drew support from many sources (Blackwood, 1984), but supporters did not disable construction equipment. Thompson states: 'this utopian ideal of young and old, conservative and anarchist, conformist and non-conformist building mutual understanding . . . didn't always happen' (Thompson, 1997: 42). Second, disillusionment with conventional politics may have led some people to direct action but a majority retained conventional political allegiances, and activists may have been attracted to a departure from the dominant society for wider and deeper reasons. I disagree with Dobson when he says, 'Any one person could be . . . a member of a green party as well as a buyer of Ecover washing up liquid' (Dobson, 1995: 145); election results show that green consumption does not translate into political allegiance. Third, the culture of the tribes, like the Dongas at Twyford Down in Hampshire, is specific to the groups involved and their historical moment. The coalitions described by Dobson did not always speak with one voice or share a radical motivation. David Pepper records 'failed marriages, loneliness and inability to afford rising house prices' more than green ideas as driving communal living in the 1980s, and views counter-cultures as 'not uninfluenced' by the dominant culture, and their values changing as it changes (Pepper, 1996: 318). I would add that the market economy is adept at subsuming departures, so that a global oil company has renamed itself Beyond Petroleum. Similarly, companies producing GM seed promote themselves as feeding the hungry rather than seeking seed monopolies. Where coalitions were effective in the 1990s was in providing information, linking self-organized groups to the media, using legal or representational means parallel to resistance, and in offering a feeling of solidarity. Fourth, by the early 1990s, activists had generated networks of mutual aid, gained skills in non-violent direct action and were mobile. In late 1990s anti-capitalism, they were skilled in using internet communications to coordinate actions and provide news services to bypass the global media industry. In the early 1990s activists took over sites and prepared them against eviction, with tree-houses and tunnels designed for medium-term occupation, as soon as road schemes were announced. Such means increased public awareness, while the state resorted to private security firms to repossess sites. All this raised the costs of road building, and in 1995 roads policy was changed. Fifth, performative irony and wit were paramount. In 1996, Reclaim the Streets (RTS) activists dressed as clowns on stilts used their voluminous skirts to hide others using road drills to dig up the M41 in west London. At Claremont Road, the most urban of anti-roads sites, artists transformed the visual appearance of the street into bricolage, drawing in a large number of visitors and tourists curious to see it. Elsewhere, the culture of music and storytelling and adoption of quasi-tribal identities contributed to a distinctive anti-roads culture, producing a sense of belonging. In a period of media spectacle, the performance of resistance played to mass audiences and provided newspapers with dramatic pictures of people in trees; but I think it was more a case that for those involved, the enactment of resistance was the new society, not its representation.

Earth First!

Through the 1980s a radical environmentalism emerged as an activist equivalent of deep ecology (see Chapter Six). The North American group Earth First! was founded in 1980 on the principles of the right of wilderness to exist, the equality of all life forms, worship of an Earth goddess, ecocentrism and the negativity of human life's impact on fragile ecologies: 'The only true test of morality is whether an action . . . benefits Earth' (Lee, 1995: 39). Similarities are evident with New Age spirituality, but the key commitment was to direct action. Earth First!, like the Weathermen

(Varon, 2004), operated as small cells, except that Earth First! used non-violent sabotage. In September 1988, an Earth First! cell sabotaged the electricity supply to the Grand Canyon Uranium Mine, putting it out of action for four days at a cost of over $200,000 (Lee, 1995: 124). The next year, the movement split, one faction adopting a millenarian call for social justice and the other prioritizing biodiversity and deep ecology. At the 1989 annual gathering – the Rendezvous – the burning of a United States flag by the millenarians renewed the split, though a new institution emerged in the form of the Earth First! Mudhead Kachinas, named after the mudhead clowns of the indigenous Zuni tribe who used clowning to release tensions. Martha Lee cites activist Mitch Friedman: 'We got naked, we rubbed mud all over, and we just started making fun of everyone' (Lee, 1995: 130). The split in Earth First! is summarized by Lee as between an apocalyptic group believing only in preserving wilderness and 'millenarians of the social justice faction [who] wanted to create a movement that could build a perfect, environmentally-sustainable society' (Lee, 1995: 139). For the apocalyptics, wilderness was a base that might survive the destruction of the present industrial society, to which they could retreat; for the millenarians, the Earth goddess was the catalyst of a transformative consciousness and interiorized route to a new society dependent also on population decrease. Chaia Heller argues that environmental crises promote an idealization of nature in the negative role of a feminine victim, while 'The fantasy of romantic protection blends perceptions of social reality with desire and fantasy' (Heller, 1999: 18), allied to a desire for purity in wilderness:

> the longing for an ecologically pure society reflects the desire to return to a time and place when society was free from the decadence associated with urban life. There is . . . depiction of the rural landscape as a vestige of a past golden age of ecological purity and morality.
>
> (Heller, 1999: 20)

Wall sees a lack of such wilderness landscapes in Britain as a reason for a rejection of deep ecology, even in the UK variant of Earth First! (EF! UK). Perhaps it is also incompatible with the social Utopia and anthropocentric tradition that characterize both the free festivals and the myth of Albion.

Reclaiming streets

Human activity and wit lie at the core of British anti-roads and anti-capitalist protest. In 1995, Reclaim the Streets (RTS) put sand on the road outside a London Underground station, bringing traffic to a halt. On the beach they had created they set out deckchairs and had a party, reminiscent of Situationism in Paris in the 1960s, and the Situationists' use of a phrase from Lefebvre: under the pavement is the beach. Lefebvre did not like his text being appropriated as a slogan, but he intended it to mean that concealed by the dulling routines of work in a capitalist society are glimpses of liberation, moments of presence that are transformative (Shields, 1999: 60–4). Rob Shields remarks of Paris in 1968: 'Beneath the surface of the alienating, rationalized city lay a ludic, festive community. As the students pulled up the granite cobbles to hurl at the police, they exposed the layer of sand on which they were laid and levelled' (Shields, 1999: 107). But, as Shields adds, the failure of the 1968 uprising devalued the currency of Lefebvre's idea of a ludic social life, and with it his critique of instrumental rationality.

In Britain, it could be argued that RTS revived the playfulness of resistance, denoting an alternative society in the ludic as non-productive, creative, non-instrumental and non-hierarchic. RTS was one of several activist groups involved in the day of action on 18 June 1999, practising dis/organization. Jordan cites an RTS leaflet:

> Reclaim the Streets London would like to emphasise that it is a non-hierarchic, leaderless, openly organized, public group. No individual 'plans' or 'masterminds' its actions and events. RTS activities are the result of voluntary, unpaid, co-operative efforts from numerous self-directed people attempting to work equally together.
> (Jordan, 2002: 69, citing www.gn.apc.org/rts/disorg.htm)

This was in part a defence against the prosecution of individuals as ringleaders, but the dynamics of mass, city-centre-wide protest extended and tested dis/organization beyond previous limits. Disparate groups improvised the day, like street theatre improvisation in extension of a commonly invested narrative. The model of dis/organization subverts the dominant society's insistence on organization, accountability, nomination, leadership, and audit. For Jordan it 'brings a little of the future into the present' (Jordan, 2002: 73). But dis/organization is not casual, and occurs only in the context of grassroots cohesion.

Through the 1990s, anti-roads protest at Claremont Road, Twyford Down and other sites gained press coverage. In East Devon, Daniel Hooper, known as Swampy, became a folk hero for his tunnelling under road-building sites (and later under a runway site at Manchester airport). But Wall comments, 'The media celebrated roads protest and crystallized the actions of thousands of grassroots campaigners into the form of a single personality. Swampy became an icon, better known, at least briefly, than many TV presenters and cabinet ministers' (Wall, 1999: 91). As Wall explains, media attention misrepresented the issue as personality when the culture of anti-roads protest was collective. For this reason I read that culture as a monastic not a sectarian piety in Szerszynski's terms (Jamison, 2001: 150; see also Chapter Six). Sectarians are counter-cultural but purposive, seeking to shift policy while witnessing against evil, but an RTS leaflet says: 'Direct action is about perceiving reality, and taking concrete action to change it yourself. It is about working collectively to sort out our own problems . . . about inspiration, about empowerment' (Jordan, 2002: 62, citing www.gn.apc.org/rts/street-politics.htm). From this it follows that activism is about consciousness, though it was also obviously about policy and witnessing against evil (roads). A particular vehicle for that emergent consciousness was – I think (as an outsider) – the tribalism of the anti-roads camp at Twyford Down.

Tribes

The tribal identity adopted by resisters at Twyford Down in the name Dongas seems to confirm the idea of a monastic piety of self-awareness. This is not the self-realization of deep ecology (Naess, 1995) and its frugality is circumstantial. Festival and fair culture is pleasure-seeking and social, while the adoption of tribalism at Twyford Down is not necessarily an appropriation of another culture but an articulation of values produced in rejection of mainstream society's institutions in an improvised, on-site culture, taking the name Dongas from an available vocabulary. McKay writes that anti-roads protest embraced issues of land ownership, environment, health, pollution, technology, business, regional and self-empowerment and self-development (McKay, 1996: 135). He records that Dongas began arriving at Twyford Down sporadically in March 1992, living on the site in an improvised camp, cooking on wood fires and building makeshift shelters. He continues:

> The term Dongas derives from the landscape the tribe came together at the camp to preserve, Dongas being a Matebele name originally adopted in the nineteenth century by Winchester College teachers for the medieval pathways that criss-cross the Downs.
>
> (McKay, 1996: 136)

The paths are old drovers' tracks. The Dongas identified with them as not-the-new-road. McKay notes that many Dongas were new to protest but disillusioned with groups such as Friends of the Earth. Informed by Earth First!, the Dongas opposed more than the government's road-building programme. McKay cites Donga Alex Plows, writing in 1995:

> Perhaps our most direct ideological links are with the environmental movement of the late sixties, seventies and eighties; for example, CND, Greenpeace, and the Greenham Women's Movement all used creative action for a global cause. Our tactics were also influenced by Earth First!; tree-sitting, climbing on machinery to stop it moving, etc.
>
> (Plows, 1995a: n.p., cited in McKay, 1996: 137)

This suggests a breadth of alternative values, while the camp offered a place in which to realize, not just pursue, them. If travelling had been a nomadic form of such realization, the camps became longer-term and large-scale moments of community. McKay views the adoption of a tribal identity as reflecting a perceived sympathy for the land in traditional societies, likening this to a similar identification at Tipi Valley in Wales. I would argue that a more prominent image of Native Americans and indigenous peoples in Asia and Australasia at the time was adopted by the campaign for land rights, and that this links to claims to the Diggers as antecedents of an incipient anti-capitalist effort to reclaim the land as common property. Plows, cited again by McKay, adds a specifically British dimension:

We are a tribe in far more than name. We have a collective purpose and a collective identity as the nomadic indigenous peoples of Britain – we have formulated our own customs, mythology, style of dress, beliefs and we are evolving our own language.

(Plows, 1995b: 26, cited in McKay, 1996: 137)

McKay explains that the Donga language consisted of New Age travelling slang – the term 'lunch out', for example, was used for someone who does not contribute to the everyday work of the camp (mainly men) – and the use of invented terms and verbal signals to identify members of the tribe approaching from a distance. The arrival of the security forces was broadcast with ululation. McKay continues that there was a clear Donga culture (in an anthropological sense), and that this is fundamental in that Dongas addressed the destructiveness of the dominant society through it, merging 'the wildly New Age' with 'the eminently and stubbornly practical' (McKay, 1996: 138).

I cite McKay extensively above because his account draws on material provided by Dongas themselves. Field says little about Twyford Down or tribalism, though he notes that anti-roads protest was not a single-issue campaign as portrayed in the media: 'the single issue is the future of human life on Earth' (Field, 1999: 75). McCreery stresses ideology, drawing on Lefebvre, Marx and Situationism, but his account relates to McKay's when he says of Claremont Road, 'every moment of every day amounted to a political act. They lived revolutionary lives, actively seeking to transform their world' (McCreery, 2001: 238); and that for Lefebvre and the Situationists, 'the key step was to begin living a richer, less alienated, more participatory culture' (McCreery, 2001: 240). Perhaps the Dongas coincidentally attempted this, drawing on an eclectic mix of cultural sources from world music and drumming to solstice celebrations. When I first looked at this material (Miles, 2004a: 213–15) I saw their tribalism as regressive, just as I still find the appropriation of indigenous cultures in the Rainbow Gatherings problematic, or at least naïve. Reconsidering the material now, I think the Dongas may have evolved a means of living and organizing close to the land, a non-violent occupation of the space between the threatened land and the state's contractors, while their improvised language is comparable to the use of identification codes in sub-cultures. Dongas tribalism may indeed be a fracture of instrumental rationality and its reliance on universal rather than material, experiential or somatic knowledges. Plows reports that after clearance from Twyford Down, 'A small group continues to live a nomadic lifestyle, moving from hillfort to hillfort with horses, donkeys and handmade carts' while others work full-time in other grassroots movements, or in permaculture (Plows, 1995b: 25–6, cited in McKay, 1996: 143).

Ecovillages

In Chapter Five I related the departure to intentional communities in the aftermath of the Summer of Love in San Francisco. Above, I suggest that ecovillages are a greening of the model of the intentional community, though there is no single model of either, and ecovillages have more than one root. There were efforts towards socially or ecologically sustainable housing in the 1970s, particularly in Scandinavia, beginning with cohousing – the first scheme being Saettedammen in Denmark, founded in 1972 (Jackson, 2002b: 157) – and in some cases leading to ecovillage settlements, such as Svanhom, founded in 1978 (Jackson and Svensson, 2002: 37; *Eurotopia*, 2000: 58). In North America, the Farm (see Chapter Five) is now identified as an ecovillage, as it uses solar cookers and showers, compost toilets, and recycled grey water for irrigation (Bang, 2005: 17–21). Many sites are in rural areas, providing for a life close to the land, or to pilot low-impact building technologies. These overlap at Brithdir Mawr, an 80-acre site in Wales established in 1995, and since 2002 combining three entities: Tir Ysbrydol, a small community with a spiritual focus; the Roundhouse Trust, which builds round huts in wood and earth (Wrench, 2000); and Brithdir Mawr Community, a housing cooperative of ten adults and five children, sharing necessary work and self-sufficient in energy and water, with community gardens and grazing land (McEvoy, 2006). There are also caring communities, retreat centres and creative arts or therapies communities. Jan Martin Bang, a resident at the Solborg Camphill Village in Norway, where carers live in an extended family structure with those for whom they care, identifies the model of an ecovillage under three headings for external systems and three for internal cohesion. He is not prescriptive, but the headings are useful in opening discussion. To paraphrase Bang, in relation to external realities, ecovillages need:

- to construct a low-impact relation to the bio-sphere;
- to construct the built environment in appropriate technologies and materials; and
- to construct an economy based on equity and fairness.

Internally, ecovillages need to consider:

- how decisions are made, how conflicts are resolved and how they relate to neighbours;
- what holds the community together ideologically, spiritually, or other ways; and
- how these questions and the factors implicit in them constitute the ecovillage as a whole.

Bang sums up this list as 'an integrated planning system' (Bang, 2005: 32), and prefaces it by a list of five general attributes. These are, again to paraphrase: human scale (defined as five to 500 members); a fully featured settlement providing all or most of the aspects of dwelling, including food, leisure and commerce; a cyclic approach to life, in which material is recycled and the settlement integrated in its environment; a supportive ambience, meeting physical, emotional, intellectual and spiritual needs; and being able to continue into an indefinite future (see Bang, 2005: 27). These points are common to many ecovillages. I would add, from my own few visits, that the most common feature is consensus decision-making, in which votes are not taken because they are divisive and meetings are frequent and sometimes long. The most common method of cultivation is permaculture (Mollison, 1988; Whitefield, 1993).

Intentionality, community, ecology

I want now to examine a few arguments and examples, loosely recasting Bang's six headings as three broad issues: intentionality, community and ecology. By intentionality I mean the factors that bring people to live in an ecovillage; by community what holds the group together, or the dynamics of eco-settlement; and by ecology the dynamics of a settlement in relation to an environment. These are headings of convenience that inter-relate and overlap. I must also add a caution that I cannot say what makes someone else move into an ecovillage. I do not live in one and have no right to speak for those who do.

Among possible factors might, however, be a desire to live close to the land or to be self-sufficient; and a willingness to share, trust, and work collaboratively with members of a group. Here there is probably some continuity between intentional communities and ecovillages. Corrine McLaughlin and Gordon Davidson (1985) note that among the intentional communities they describe, mainly in North America in the 1980s, most had shared spaces for eating and various kinds of work from laundry to vegetable and fruit cultivation, and that many shared child rearing as well. Drawing on the experiences of several long-established communities, they relate that while privacy is an issue, people develop coping tactics and find it less important after a period of community living (McLaughlin and Davidson, 1985: 67, 367 n.12). The intention, then, of the intentional community is to find self-organizing ways to negotiate such matters. In ecovillages the same tends to apply, with an emphasis on the ecology of human settlement. McLaughlin and Davidson identify different kinds of intentionality, however, from the goal-oriented intention of Arcosanti (see Case Two), aiming to become a city of 5,000 people with a building programme, technology and pre-determined architectural style, to the process-based intentions of the Alcyone Community in Oregon, where members spent more than a year 'working out their interpersonal problems with each other before trying to build anything on their land' (McLaughlin and Davidson, 1985: 73–4). In a few days at ZEGG (Zentrum für Experimentelle GesellschaftsGestaltung – Centre for Experimental Culture Design), an ecovillage in Germany (Case Nine), I realized that expressing feelings about others is vital to the sustainability of a settlement, and that meetings are less about an agenda than a space in which to express feelings and so nurture a community. Because these processes are important, often underpinning the intention of the settlement, new members are generally invited to make preliminary visits and undergo an induction period before any decision to join a community (or be accepted into it) is made. An ecovillage could be seen, then, as a human-scale community in which the inter-relation and negotiated interdependence of each individual defines the ethos of the community. Karen Svensson writes:

> Ecovillages are communities of people who strive to lead a sustainable lifestyle in harmony with each other, other living beings and the Earth. Their purpose is

to combine a supportive social-cultural environment with a low-impact lifestyle.
. . .

> The deep motivation for ecovillages . . . is the need to reverse the gradual disintegration of supportive social-cultural structures and the upsurge of destructive environmental practices on our planet. Underlying the concept . . . is the desire to take responsibility for one's own life; to create a future which . . . is regenerative for the individual and for nature . . . A future which we would like to bequeath our children . . .
> Ecovillages are a way of living. They are grounded in the deep understanding that all things and all creatures are interconnected, and that our thoughts and actions have an impact on our environment.
> (Svensson, 2002: 10)

Low-impact living is also practical, at times pragmatic, and not only ideological. A desire to be self-sufficient in energy or food, outside the fossil-fuel and agri-business economies, needs land. For ZEGG, land in the former East Germany was relatively inexpensive, which drew an existing alternative group to go there. On a map of ecovillages and intentional communities in Germany in 2000, around one-third of the one hundred or so shown are in the east (*Eurotopia*, 2000: 88–9). Several others re-use redundant industrial buildings, as at Lebensgarten in a 1930s munitions factory. Lebensgarten houses around 90 adults and 50 children, offering a programme of workshops on personal development, conflict resolution, health, meditation and ecology (*Eurotopia*, 2000: 138).

The relation to the land varies between food production and wilderness, with land restoration somewhere between the two. The latter may entail re-adapting an ecology after human exploitation. For example, Alan Watson Featherstone of the Findhorm Community in Scotland writes of the replanting of trees in Glen Affric:

> Under natural circumstances, the trees in the forest would regenerate by themselves, but the artificially high grazing levels maintained by human interests prevent this. So when we plant trees, we do so in patterns which copy the distribution of naturally regenerating tree seedlings in the glen, and in the same soil types and topography as the trees grow in by themselves.
> (Featherstone, 2002: 27)

This is similar to the land restoration programme, since the 1970s, at Auroville (Case Three), where seeds were obtained from local temple gardens in which localized plant ecologies were protected. At Ithaca ecovillage (Case Eight) people seek to preserve the surrounding woodland for birdwatching or moon walks, but the relation to the land is also productive in permaculture gardens and polytunnels. Hildur Jackson writes:

> People's wish to produce their own food springs from a basic human need to connect with nature, to provide for oneself, and to play an active, responsible

role in one's natural environment. This need can be met in various ways, by cultivating one's own gardens (including rooftops), sharing a plot of land with others, or also growing vegetables and fruit in an ecovillage/community setting.
(Jackson, 2002a: 31)

Something similar might be said of an urban allotment. Geographer David Crouch writes of the creativity of allotments, and the informally regulated community: 'Getting on with each other is important . . . because the margins of what any one individual does are not clearly defined. Many plotholders don't easily get on . . . [But] a prevailing if perhaps exaggerated theme amongst allotment holders is the significance of friendship' (Crouch, 2003: 15). Jackson goes on to say that local food production frees people from dependence on agribusiness, or from cash crops for export in the non-affluent world. She asserts that bio-regions should seek self-sufficiency in food production. I find this complex, first given the growth of fair trade – which means not all imports are exploitative, though fair trade may mask inappropriate, monocultural land use – and second in that self-sufficiency eliminates trade as a means of contact that might have other values. On the other hand, the model of a city self-sufficient in most of its food would be interesting and contribute to a large reduction in the embodied energy of food (its transport, packaging and so forth). Helena Norberg-Hodge writes that community supported agriculture (CSA), which includes vegetable box delivery networks, subverts the global food industry and a centralization of economic power, and describes a growing tendency to support local economies: 'Within this countercurrent, attempts to link farmers and consumers are of the greatest significance. A local food movement is sweeping across the world', its benefits including rebuilding local economies, fostering diversity, and protecting natural resources (Norberg-Hodge, 2002: 39). Cambridge Cohousing (Case Seven) has a link to a regional CSA scheme. But is it possible to go further, to regenerate local food production and develop ways of living that avoid the use of fossil fuels?

Back to the land

Tinker's Bubble is a settlement of 10 adults and two children in rural Somerset, a 40-acre plot with 27 acres of woodland bought by the group which established the settlement, at first as a camp, in 1994. It uses no fossil fuels. There are common and individual plots so that dwellers can make a living from the land. A monthly meeting decides policies and actions. Each person contributes one day a week to communal work, which, with a horse (called Samson), is enough to manage the common land and woodland. There are two cows. Each person pays an agreed sum (£12 per week in 2004) to cover bought-in food, tools, insurance, Soil Association affiliation and loan repayments. The group intends to remain small, but was initially joined by anti-roads activists. Simon Fairlie, one of the initial members, has built up several years of experience dealing with the local authority planning system, and now advises other groups on low-impact dwelling. The system generally prevents building even low-impact and sustainable dwellings on agricultural land (while gaining planning permission for a plot in a village increases the land value many times, making it unaffordable for rural workers). After a public inquiry resulting in refusal of permission, and losing an appeal, the group eventually received permission for up to 12 low-impact dwellings and three common buildings for five years, after which a reapplication is necessary. Fairlie accepts this outcome because the community was still establishing itself at the time and not ready to build permanent structures. The first few years, undergoing the planning ordeal, were spent living in tents, yurts and makeshift structures. Current plans are for a group of wooden structures. Douglas Firs in the woodland are suitable for self-build timber houses, and there is a horse-powered saw mill on site. By 2004, a house and stable and a common roundhouse (actually octagonal) had been built. The woods can sustain production of 90 tons of timber annually, using the horse to bring it out of the woods. In addition there are traditional apple trees.

Dorothy Schwartz visited Tinker's Bubble in 1998, writing that she met a 19-year-old anti-roads protester whom she quotes as saying that she is 'not into farming but protesting. And besides my parents wouldn't approve' (Schwartz and Schwartz, 1998: 48). Another dweller says she does not seek publicity, adding, 'how would you like someone asking questions while you are having your breakfast?' and continues: 'My days here make me question our own work ethic. Much of what we consider housework does not exist here – no furniture to dust, beds to make, surroundings to prettify. Does that leave everyone longer time for meaningful discussions or purposeful activity? It does not seem to' (ibid.). Tamsin Blanchard, a journalist writing in 2000, interviewed an ex-teacher and forestry student who had built a shelter in willow, ash and hazel, covered with old carpets under a tarpaulin. She notes the tidiness of the makeshift home, a fan of peacock's feathers, and that 'the children's rooms are packed with books and toys, just as you'd expect' (Blanchard, 2000: 59). Blanchard compares this 'ideal home' with others in the nearest village, and begins her account of the home she visited in Tinker's Bubble by pointing out that it has 'three bedrooms, two toilets, an open plan dining/kitchen/summer room, workshop, extensive gardens . . . [and] some of the most magnificent views in all of England, looking out across the rolling green countryside' (Blanchard, 2000: 56). Emma Chapman writes of Tinker's Bubble in these terms:

> The unobtrusive, temporary dwellings; the modest incomes gleaned from a range of sources; the direct reliance on their land and themselves (water from a spring, firewood grown on site, compost toilets, owner-built homes); the environmentally-sensitive land use – all these make Tinker's Bubble an example of low impact living.
>
> (Chapman, 1998: 21)

Tinker's Bubble became known as a test case for planning permission for low-impact living, but I have not seen it referred to as an ecovillage. Perhaps it is instructive that, outside the ecovillage movement, there are other models of settlement in which living close to and caring for the land are combined with negotiated individual autonomy and interdependence. But Tinker's Bubble is also based implicitly on principles of social justice. As I understand it from a meeting with Fairlie and the published sources cited above (see also Fairlie, 1998), Tinker's Bubble is a demonstration of both low-impact living and a claim for a right to live on the land for those who are not members of the established property-owning class – as was the underlying purpose of the Charterville Allotments (Chapter Three). In this context, Chapman ends her article describing the 'tiny footpath winding uphill' to the settlement, and refers several times to the importance of it being out of sight as a way to ease its path through the planning process; then she asks, 'Have they the right to live in Britain's precious, dwindling countryside?' (Chapman, 1998: 22).

This links, too, to the efforts for autonomy of plotland dwellers in the 1920s and 1930s, who underwent similar battles with planners and landowners. Dennis Hardy and Colin Ward write:

> Against a wider background of property and control, nowhere were the issues more polarised than in the plotlands. Here, on one side, were people of modest means defending pathetic homes on land that none had previously wanted . . . and, on the other side, a bureaucratic machine without compassion, growing with every new bye-law and Parliamentary clause. . . . Plotlanders did not know that they were in the front line of a battle that had little to do with their immediate needs, while conscientious local authority officials in pursuit of modest improvements in public health were frequently perplexed by the weight of an opposition which seemed far removed from reason and fact.
>
> (Hardy and Ward, 2004: 33)

Jeffrey Jacob, writing of the North American back-to-the-land movement, argues from social research among ex-urban rural settlers in the 1980s and 1990s that the values articulated by his respondents included peace of mind, union with nature, joy, living in the moment, wonder, wholeness, and a sense of being accepted in the universe (Jacob, 1997: 85), while their satisfaction was high in respect of family life, housing, relations with neighbours and self-reliance, among other factors (Jacob, 1997: 100). Jacob writes of their motivations as modest, though dealings with bureaucracy can turn people to activism – 93 per cent of Jacob's respondents were sympathetic to activism, and 41 per cent belonged to an activist group (Jacob, 1997: 179). Questions of a right to the land, low-impact living and joy – as moments of presence amid routine – draw together issues of social justice, environmental sustainability and liberation.

There are also issues of environmental justice, particularly in North America where pollution affects poor people and people of colour disproportionately. This produced a new activism, focused on urban, industrial sites and drawing on the experiences of both civil rights workers and local campaigns in the non-affluent world. Steve Lerner writes of a new wave of community organizers 'who call themselves eco-justice activists', who work 'in communities of colour and impoverished areas . . . organizing local grassroots movements to protect public health and improve living conditions' (Lerner, 1998: 321). Natural building, too, has been linked to opposition to the dominant society's political and economic structures, not just to sustainability, and Robert Bolman writes, 'We must consider social realities when we contemplate building our houses' (Bolman, 2002: 14). He continues that natural building – in earth or straw bale, for example – addresses the sickness of consumerism, and that 'The natural building movement exists hand in glove with the "voluntary simplicity" movement' (Bolman, 2002: 15). Meanwhile, at the time of writing, the roundhouses at Brithdir Mawr are under threat from the planning system that Fairlie sees as threatening nearly all low-impact and permaculture sites in Britain (Fairlie, 2003). This denotes a form of inverse institutionalisation of alternative dwelling, as it deals with bureaucracies about which it must therefore be informed if it is to have any prospect of gaining the desired right to live on the land. Andrew Jamison writes:

> Around 1987 the environmental movement began to take on a new character. Metaphorically we might say the seeds that had been planted in the cultural soil had started to take root. The utopian practices which had characterized much of the activity of environmental movements up until then have gradually given way to the somewhat more complicated and diffuse work of instituting and implementing: cultivating and nurturing the seeds . . . As a result the underlying meanings of environmental politics have come to be fundamentally altered. Other interests have had to be taken into account . . . The terms of discussion have shifted from making particular improvements . . . to integrating environmental concerns into all other kinds of social, economic, and political activities.
>
> (Jamison, 2001: 176)

Perhaps this is how activism comes of age, renewing the coalitions of anti-roads protest in interactions with the dominant society's legal and political apparatus.

Crystal Waters

The extent to which coalitions and alliances operate among elements of the emergent alternative society may be a factor in its future growth and ability to negotiate a future within mainstream society and its planning systems. This would constitute a long, or incremental, revolution; but it would also constitute a new kind of millenarianism inasmuch as the state of being of those living in alternative settlements is itself revolutionary and at times joyful (Jacob, 1997). In that respect, living on the land might be a monastic piety in Szerszynski's terms (see Chapter Six), a principled rather than purposive knowledge, even though it needs to be purposive in dealing with the planning system. But I want to end this chapter by looking briefly at permaculture as an example – by no means the only one – of practical sustainability.

Permaculture translates a holistic approach to life into the design of cultivation for minimum invasiveness and maximum mutual benefit of species, so that plants, trees and crops, and any animals and birds husbanded on permaculture plots, all contribute to the conditions required by other elements for growth. This is not wilderness culture but its opposite – design, or a land planning process. Permaculture orders the elements of the natural world in a way deemed to be conducive to their flourishing, but is also a making new, often in spaces of previous neglect (though it is used in suburban gardens as well). Permaculture is thus improvement, but may not carry the burden of reform in the sense of nineteenth-century liberalism, where improvement was undertaken by an elite as a means to prevent insurrection among the poor. Permaculture tends to be carried out by ecovillage communities or groups working specifically on land issues. Bang envisages a spectrum with permaculture at one polarity and agribusiness at the other, one close to natural growth, the other using chemicals to produce higher yields at the price of soil degradation. He cites hydroponics – a system of growing plants in gravel while nutrient-rich liquid is passed constantly through it in a controlled system – as at the opposite polarity from permaculture (though it is used in a few cases of green agriculture – see Bang, 2005: 131). Organic farming is towards the middle, but can be open to an equivalent of industrialization. Permaculture can be applied in urban roof gardens, and is, he says, 'not just about having a subtropical plot and scattering seeds about and harvesting haphazard results. It's about planning real life solutions to real life situations and helping more people gain power over their own lives' (Bang, 2005: 131).

To me the idea of permaculture loosely resembles the anarchist principle of mutual aid (Kropotkin, 1902). The actions, so to speak, of plants and trees are not voluntary in the human (or arguably animal) sense, and I do not want to talk of plants helping each other as if to anthropomorphize them. Nonetheless, if permaculture remains a human preserve empathetic with natural growth, it rests on mapping the human principle of cooperation, and working at the tasks one is best able to accomplish, onto the relations between different species in a material context. It could serve as the model for human social organization, in which the actors become human actors intervening in their own ordering, making synergies and gravitating to others with whom they can co-prosper – rather as proposed by Fourier (see Chapter Three). Permaculture is used at ZEGG as integral to a way of dwelling based on trust. Its best-known site, however, is Crystal Waters in Queensland, Australia.

Walter and Dorothy Schwartz describe their visit to Crystal Waters, surprised that it looks like 'a manicured garden suburb' instead of meeting their expectation of a 'jungly paradise on the wet, sub-tropical Queensland soil which can turn a gum sapling into a gum tree in a couple of summers' (Schwartz and Schwartz, 1998: 127). They note the absence of a village centre, and a variety of house designs in earth, pole-frame and other materials. They continue: 'the site looked so prosperous, so comfortable. Would this turn out to be little more than an original hideaway for the sixties baby-boomers who had flirted with the age of Aquarius?' (ibid.). The geodesic domes of some of its buildings might add to that 1960s feeling, but Crystal Waters was established as a permaculture village in 1987, with roots in the back-to-the-land movement of the 1970s. It occupies a 259-hectare property (83 separate lots), previously used by indigenous women for initiation rites. In 1979 it became the site of a commune, but by 1984 internal divisions led a number of people to leave, at which point a plan by Max Lindegger, Robert Tap, Geoff Young and Barry Goodman for a permaculture site was adopted. Fifty thousand trees have been planted, and there are ponds and fields of grain and sunflowers. The 250 or so residents are autonomous on their own plots, subject to collective rules only in relation to the common land and certain principles: there are no dogs or cats (in the face of a national problem of feral animals); chemicals are not used on the land (to retain organic status); there is no fishing; firearms are not allowed; and residents must not annoy neighbours. One-third of the common land is used for income generation from horses, lotus roots, forestry and pecan and nut plantations, to provide a basic income. The Schwartzes write of conflict over the lotus root project, some residents being anxious that the species will colonize nearby natural sites, or that the plants will fill the man-made pond in which they grow to deprive people of an amenity. The Schwartzes add:

> Management of Crystal Waters has run into other difficulties too. A levy on all members is supposed to pay for common services like fencing and roadside trees. The trouble is that many members are cash-poor, quite a few live on the dole, and they always oppose such levies. There is a shortage of ambitious projects because of the law which says all communal ventures must be agreed unanimously. Would-be entrepreneurs find it hard to get loans for commercial ventures with so shaky a legal title. . . .
>   A more fundamental question: is Crystal Waters succeeding as a community? Some members find the long, boomerang-shaped layout unsociable . . . Other critics find fault with the design itself, complaining that permaculture theorists have been insensitive to community needs.
> (Schwartz, 1998: 135)

But not everyone there practices permaculture. Others were attracted to a minimum of routine and a sense of community. Still, when the Schwartzes say that Crystal Waters 'addresses more completely the physical environment than the social' (Schwartz and Schwartz, 1998: 141), I think of Garden City (Chapter Four) as a design solution to social and environmental needs that lacks the means to produce a sense of community. But a sense of community can be produced only by those who are the community, in everyday life and in the kinds of interactive processes through which they reach consensus, draw out negative as well as positive energies among the group, and learn to live together as a self-organizing and constructively self-critical society within the dominant society. I wonder whether perhaps when a group reaches a certain size, maybe more than 100 or so, this becomes difficult. A difficulty at Auroville, for example, seems to be that only at most one-quarter of the city's inhabitants participate in its open meetings. On the other hand, managing a meeting of Auroville's 1,700 people of all ages might be problematic, even though a suitable site is provided. A challenge for the growing alternative society appears to be, then, how the research and development work of self-organization that takes place in ecovillages can be adapted for mainstream, urban societies. Having said that, the experiences of ecovillage dwellers tends to suggest that social architectures are more important in sustaining a social formation than built architectures. I wonder if in a similar way the transformative consciousness of activism and ecovillage living underpins the sustainability of alternative societies.

Built and social architectures

I am conscious of my outsider status in relation to activism and ecovillage living, but end with a few observations. First, the ethos of an alternative society is articulated and continuously reshaped in dialogic processes. This seldom happens in the atomism of mainstream society. Second, the tools of the mainstream society are unlikely to dismantle its codes (see Lorde, 2000), and there are questions as to how an agreed way of living is arrived at, and as to its flexibility. A common criticism of Utopia is that it is rigid and unable to accommodate change. Third, the built architectures of a settlement may be conducive to community cohesion and personal satisfaction, or aesthetically pleasing, but in many cases the buildings are recycled and work just as well, while there is no clear correlation between the joy of those who live in communities occupying new, purpose-design eco-homes and those who live in recycled sites. Again, there are technologies and materials of natural building (Kennedy et al., 2002) but these do not guarantee that a settlement will be socially and culturally sustainable. Some such methods draw on indigenous traditions, or adapt and seek to improve a vernacular; others use a sense of local scale and patterns of using space, but in a mix of new and old materials and structures. An example of the latter is the work of the Rural Studio established in the southern United States by Samuel Mockbee (Dean and Hursley, 2002), producing low-budget, low-impact houses for poor people in rural areas. The Rural Studio did not design ecovillages, nor intervene in social processes outside those relating to the provision of houses and local amenities within the planning and education systems (Mockbee being a teacher at an architecture school). At the same time, it may be that its concern for social equity is, like grassroots work in the non-affluent world, a vital contribution to a culturally and socially sustainable world.

The problem, which I do not see as a difficulty (in the sense of a flaw in an argument), is how to map the experience of the non-affluent world onto the affluent world, and how to make the necessary and consequent changes in the role and status of professionals, and to equate expertise with the tacit, local or barefoot knowledges of dwellers (some of which I describe in Chapters Eight and Nine).

In the end, the social architectures of alternative societies appear more developed than the equivalent built architectures. In this respect, the example of ZEGG (Case Nine) is interesting, its vision described by Achim Ecker as 'the realization of inner and outer peace and a sustainable lifestyle that is not based on the exploitation of nature and humans in other parts of the world. The common thread among us is the quest for new forms of love, sexuality and the manifestation of inner and outer peace' (Ecker, 2004: 27). I discuss ZEGG below but will end this chapter with another quote from Ecker:

Communities ... show that sustainability leads to a higher quality of life. Our aim is to combine personal growth, the implementation of new ways of living together, and a political commitment towards the world. We believe that we need to work both on the inner, more personal, and the outer, more political, level in order to achieve the world we would like to live in and pass down to our children.

(Ecker, 2004: 30)

Part Three
Horizons

8 Mud-brick Utopias

Introduction

In the previous four chapters I considered the utopianism of Modernist architecture and planning in the twentieth century; the moment of the Summer of Love and growth of intentional communities in the 1960s; approaches to environmentalism; and the growth of environmental activism and ecovillage living in the 1990s. I noted the importance of grassroots activism in the non-affluent world in Chapter Six, and turn now to focus on the non-affluent world in this chapter and Chapter Nine. In this chapter I investigate the work of Egyptian architect Hassan Fathy, looking mainly at two pilot projects using mud-brick building: the village of New Gourna (1945–9) and the cooperative settlement of New Baris (1965–7). I situate Fathy's work in relation to traditionalism and nationalism, and argue that it constitutes an alternative Modernism as utopian as Western Modernism. I ask next whether Fathy's utopianism is also a form of social engineering, and whether the fusion of design and construction, and the potential participation of dwellers in the making of mud-brick buildings, offers precedents from which the affluent world might learn today. Finally, widening the chapter's geographical scope, I cite a small number of housing projects in the non-affluent world including Balkrishna Doshi's work at Aranya, India in the 1980s. In Chapter Nine, I turn to the work of the Social Work Research Centre, a rural campus at Tilonia, Rajasthan, known as the Barefoot College.

Pasts and futures

Fathy's first major project was the planning, design and supervision of construction of a new village in the agricultural zone on the west bank of the Nile, between the river and the archaic monuments of the Theban necropolis. The purpose was to rehouse the inhabitants of Gourna al-Jadida, the old village in the archaeological zone. The project began in 1945 and was abandoned four years later when most of the public buildings but only one-fifth of the housing in the plan had been constructed. Most villagers then refused to move, though New Gourna is fully occupied today. Some houses have been modified, others are as Fathy left them, though there is some deterioration. Most Gournis still live in the old village – a series of five hamlets on the side of a limestone ridge beneath the Pharaonic Tombs of the Nobles – where they produce ceramic and alabaster artefacts, are employed as labourers or guardians in archaeological sites, act as guides, or improvise a livelihood from multiple sources mainly aligned to tourism. When a storm damaged a number of houses in the old village in the mid-1990s some families were forcibly moved to a second new village – a government housing scheme of two-storey brick and concrete blocks in straight lines at the edge of the desert. I visited the site in 1996. Naïvely asking the taxi driver why people moved there, I was told 'because they shouted at them'.

In the West, Fathy's work has received renewed attention for the low environmental impact of mud-brick building, seen in context of the tradition of adobe in the Americas (Doughty, 2002). But while the choice of adobe in Taos, New Mexico, a node of affluent alternative lifestyles, is a matter of self-image, for Fathy it was a way to provide a 'no-cost' architecture for a developing country (Fathy, 1988: 16). Mud is free, and the labour of dwellers cooperating in the building of their own homes is a means to social cohesion. Fathy also saw mud-brick building as a means to regenerate village society and decrease the influx of the rural population to Cairo. The social architecture of mud-brick building tends, however, to be overlooked in its aesthetic appreciation. Fathy's book, *Gourna: A Tale of Two Villages* (1969), better known under the title of its second edition, published in the West as *Architecture for the Poor* (1973), is frequently cited, but there is more than one possible context for this: the aesthetics of earth architecture (Kennedy *et al.*, 2002; Goldfinger, 1993; Dethier, 1982); interest in emerging national cultural identities and indigenous modernities (Hosagrahar, 2005); and concern for the economic, social and cultural stability of informal settlements after John Turner's research in Latin America in the 1970s, and growth in alternatives to Western development (Serageldin, 1997; Hamdi, 1995; Turner, 1976). Some informal settlements have been legalized and accepted as viable forms of settlement (Fernandes and Varley, 1998), characterized, as Dirk Berg Schlosser writes, by 'internal communal feelings' (Berg-Schlosser, 2003: 218). Having said that, Gourna-al-Jadida is not an informal but a traditional settlement, and New Gourna is a government housing project, its aim to improve the living conditions of villagers while changing their pattern of economic life to suit government policies.

New Gourna arose in 1945 from efforts to preserve the Pharaonic past on the part of the Egyptian government's Department of Antiquities. The Gournis were expert at finding tombs and had in many cases built their houses over tombs. In the 1930s and 1940s they were employed as labourers on archaeological digs, and also as draughtsman recording the interior decoration of the tombs. The Gournis depended on such work but had their own informal economy. While the colonial powers freely expropriated Egyptian antiquities to stock their museums, following the Napoleonic invasion of Egypt in 1798, found artefacts were also traded by villagers with foreigners eager to acquire them. For the most part these were minor finds but to the authorities the trade constituted theft. Fathy notes that the Gournis had exhausted the richest treasure 'long before the antiquities fetched a really high price' (Fathy, 1969: 15), and himself incorporated a Pharaonic relief in the interior courtyard of a house he designed for Hamed Said and his wife, both artists, as a weekend retreat near Cairo (Steele, 1997: 54–9; Richards *et al.*: 40–1). Matters came to a head when a major artefact disappeared. A decision was taken to move the Gournis to a site away from the tombs. A 50-acre site was compulsorily purchased on agricultural land nearer the Nile for a new village. From the beginning, then, New Gourna represented a forced relocation for the 7,000 or so villagers. Fathy sought to redress this imposition by providing a better environment in the new village, analysing the spatial practices of the old village in order to replicate them in the new plan. Fathy's preference for local over imported technologies alienated officials in Cairo, whose interests were not served by vernacularism, while almost all the villagers refused to be moved. This does not detract, however, from the intrinsic interest of the project as a renewal of a vernacular process of building in which the architect's role becomes that of planner and facilitator. Among the difficulties are that Fathy was inevitably complicit in the authorities' aims, and that his attitude to the poor, while beneficent, is coloured by his status as an urban professional.

Fathy writes of a visit at the beginning of his career to a small country town, Talkha, where he saw 'mud and every kind of filth . . . squalid little shops, fronts open to the smell and flies in the street, displaying their few wretched wares to the poverty-stricken passersby', adding that the peasants are hopelessly resigned to such conditions – 'too sunk in their misery to initiate change' (Fathy, 1969: 3). There is, too, an element of eclecticism in Fathy's vernacular, which uses the material common to all Egypt's rural areas (and at that time the poorer parts of towns as well) but does so in a style and building technique drawing on the types and skills not of the area around Luxor but from Nubia, three hours by rail to the south, and from medieval houses in Cairo. Fathy's rhetoric at times implies pursuit of a regressive traditionalism that could also be read as an urban attitude to a mythicized countryside (at odds with his perception of the provinces cited above). Yet, despite all that, I argue that it is equally a progressive utopianism and that this is in keeping with the invention of a vernacular rather than its reproduction.

National culture

If the first context for New Gourna, and Fathy's revival of mud-brick building in general, is the relocation of the Gournis, together with his ambivalent view of conditions in rural Egypt (which I elaborate below), the second is Egyptian nationalism, leading up to the revolution of 1952. A growth of national sentiment among Egypt's professional and commercial class in the 1940s engendered a search for a national cultural identity. Egypt was seen externally as an archaeological site, and if it had other identities these were as part of the Ottoman Empire and site of the Suez Canal, or as a tourist destination. Egypt figured in the Orientalist paintings of John Frederick Lewis, Thomas Seddon, Ludwig Deutsche and Jean-Leon Gerome, for example, in the early and mid-nineteenth century, and was visited by Gustave Flaubert (Steegmuller, 1983) and Gérard de Nerval (Nerval, 1972), both of whom wrote accounts of exotic sexual encounters. In contrast, Amelia Edwards published an account mainly concerned with ruins, from a journey to Khartoum in 1873 (Edwards, 1982) – five years after Thomas Cook first conducted white people to Aswan. Lewis may have lived 'the life of a lotus eater in Cairo'

(Jullian, 1977: 134) in the 1840s, but Orientalism misrepresented the near East as a site of Western desires, and was of no interest to Egypt's professional elite. Edward Said writes that Orientalism shares with myth and magic 'the self-containing, self-reinforcing character of a closed system, in which objects are what they are *because* they are what they are, for once, for all time, for ontological reasons' (Said, 1985: 70). But national cultures tend to epistemology. The Pharaonic past, too, was revered by foreigners who saw it as a dawn of civilization, but of *their* civilization; while for Egyptians it was the remains of an ethnically different society. Ancient Egypt was expropriated in museum holdings and colonial narratives but closed to reappropriation in a nationalist (rather than merely national) cultural identity.

There were only two other possibilities: an Arabist style, aligned to an international Islamic culture, uniting its public in a common faith; or a folk-realist style based on but reinterpreting village traditions. In the 1940s, among the urban professional and artistic circles to which Fathy and the Saids belonged, the folk-realists were the most influential group (Karnouk, 1988: 47). Rural life was as remote to these urbane professionals as it was to government officials, but the folk style enabled a reclamation of an identity outside colonialism, which was also outside the vocabularies of Western Modernism. The work of folk-realist artists was shown in galleries patronized by urban bourgeois collectors, but as a symbol of an emergent national independence movement. This appears to me an important aspect of Fathy's vernacular revival, perhaps more so than his complicity in a forced relocation of the Gournis (which in any case failed). Frantz Fanon writes of the tendency of colonized people to look to a remote past for legitimation of the freedom to which they aspire:

> the past existence of an Aztec civilization does not change anything very much in the diet of the Mexican peasant of today. . . . But . . . this passionate search for a national culture . . . finds its legitimate reason in the anxiety shared by native intellectuals to shrink away from that Western culture in which they all risk being swamped.
>
> (Fanon, 1967: 168–9)

And it may be that folk-realism carried this role. Fathy sought to revolutionize the folk tradition, then, by improving the conditions of the poor in a new vernacular. This means that Fathy's traditionalism may not affirm traditional moral and social conventions in a generally conservative culture, either, but is a departure from the aesthetic colonialism of European-inspired taste – the taste that tinted perceptions of Egypt as an exotic site. It is, too, and importantly, a practical refusal of the economic colonialism effected by imports of inappropriate and expensive European technologies and materials of building.

A romantic rural idyll?

Fathy contrasts hand-made craft goods, with their irregularities and adaptations arising in the process of manufacture, to the standardization of machine-made goods, writing that mass communications, mass production and mass education characterize both capitalist and communist societies (Fathy, 1969: 27). But his description of his arrival at Luxor by sleeper train prior to beginning work at Gourna in 1945 indicates the extent to which he was an urbane professional:

> I got off with all my suitcases, trunks, rolls and rolls of plans, instruments, gramophone, records, bits, pieces, odds and ends . . . and I found a great crowd assembled to meet me. This crowd was composed of all sorts of people who had some connection to the work or who hoped to be engaged, and with it I set off like a sultan for Gourna.
>
> (Fathy, 1969: 152)

James Steele argues that Fathy draws selectively on European Enlightenment ideas, and that his viewpoint was derived from 'models specifically indicated in the *Description de l'Egypte*' (Steele, 1997: 184). The *Description*, produced in Paris in 23 volumes between 1809 and 1823, used material recorded during Napoleon's 1798 expedition. Said writes that it provided a context as well as a geographical setting for Orientalism, 'since Egypt and subsequently other Islamic lands were viewed as the live province . . . of effective Western knowledge about the Orient' (Said, 1985: 41–2). But it is against this that he reacts, though Fathy's reasons are personal as well as public.

At the beginning of his book on New Gourna, Fathy writes of his family background: 'My father avoided the country. To him it was a place full of flies, mosquitoes, and polluted water, and he forbade his children to have anything to do with it' (Fathy, 1969: 1). As it happens, in 1943–4 a malaria epidemic killed one-third of the population of Gourna. Fathy notes this, and (in the context of a cholera outbreak in lower Egypt) the fact that villagers drank water from open wells (Fathy, 1969: 166). But he also writes:

> My mother had spent part of her childhood in the country, of which she preserved the pleasantest memories, and to which she longed to return right up to the end of her life. She told us stories of the tame lambs that would follow her about . . . the chickens and pigeons, of how she made friends with them and watched them through the year. . . . She told us how the people produced everything they needed for themselves in the country, how they never needed to buy anything more than the cloth for their clothes, how even the rushes for their brooms grew along the ditches in the farm. I seemed to inherit my mother's unfulfilled longing to go back to the country, which I thought offered a simpler, happier, and less anxious life than the city could.
>
> (Fathy, 1969: 1–2)

Fathy's picture of village life combines his father's negative perception with his mother's nostalgic aspiration. He describes a peasant house as holding 'a large variety of bulky stores and the owner's cattle as well . . . hens running in and out among the dust and babies' (Fathy, 1969: 92). This (though based on lower Egypt, the Nile Delta) is less sympathetic than expatriate Winifred Blackman's description of village houses in upper Egypt twenty years earlier:

> In the better houses there is generally a flight of steps leading to an upper storey, where there may be a sitting room . . . The flat roof is a pleasant place on which to sit and watch the life in the streets below.
>
> (Blackman, 1927: 27)

From the negative image of the countryside he received from his father, then, combined with the positive, if romanticized, image bequeathed him by his mother, Fathy introduces an ideal of village life derived from several sources. This, already an eclectic vision, is enhanced by details from houses in the old quarters of Cairo and from building types in other parts of North Africa that Fathy visited at different stages of his career.

New Gourna

When he was appointed as architect responsible for New Gourna, Fathy was not well established, having taught at the Fine Arts Academy in Cairo but with few buildings to his name. His first project was a house for the Egyptian Royal Society of Agriculture at Bahtim in 1941, destroyed soon after construction on the grounds that the Society had its own architect with whose work Fathy's use of mud-brick did not fit. Fathy designed the Said House a year later: it was enlarged in 1945 and used as a meeting place for the Friends of Art and Life, a group including architect Ramsis Wissa Wassif, founder of an art centre at Haranya, near Cairo (Steele, 1997: 58–9). Fathy also designed houses to replace those lost in a flash flood at Ezbet Al-Basry, villas for elite clients, and then, in 1945, a house for Fawzi-Bey Kalinia at Melia which used a wind-catch (*malkaf*) and lattice-work screen (*mashrabiya*) – details from houses in Cairo (Steele, 1997: 36–9). New Gourna was an opportunity to test his approach on a social scale.

The site of New Gourna has few advantages. Set amid fields of sugar cane, bounded by a light railway used to collect the harvest, it is too far from the monuments to provide easy contact with the Gournis' traditional source of livelihood. There are no tombs in the mud, nor artefacts to be found in it. Nor was the project budget adequate. Fathy writes that 50,000 Egyptian pounds was allocated nominally as compensation for the villagers, but 'This money was to be turned over to me to build a complete village of nearly one thousand houses' (Fathy, 1969: 69). He records that, using his previously tested mud-brick method, it was enough for the houses but made no allowance for roads, schools, a mosque or public buildings. Fathy was allocated six assistants to help him oversee the project, which required a workforce of 300 labourers to be recruited locally, but their qualifications were in art and craft, not architecture or engineering. He adds:

> I reflected that it was not really so important that I had no supervisors to help me. The important thing was building, and that would be done by the Aswani

masons. They would work without supervision and could, indeed, teach a thing or two even to qualified architects.

(Fathy, 1969: 151)

The recruitment of a workforce was no easier. Fathy wrote to the sheikhs of each hamlet asking for a list of people suitable as labourers. None replied. Meeting one of the sheiks, Fathy records being offered a precious stone as a bribe, followed by a cautionary tale of an official recently dismissed for failing to heed local advice. Offered food, he writes:

It was a pie, a huge, soggy pie, which gave me food poisoning just to look at it. Through my mind went all the stories I had heard of peasant pride ... I thought of my standing among the Gournis; and I decided. I stood up and swore ... that I would not touch a crumb of his nasty pie.

(Fathy, 1969: 159)

Coffee arrived in a dirty cup, at which point Fathy left, resolving to design a hygiene centre in the new village where women could attend cookery classes. These anecdotes denote his distance from local culture but the villagers, it should be noted, had a long history of dealings with outsiders seeking their compliance in matters not usually to their advantage. Workers were recruited in due course but when they tried to dig sand to mix with mud and straw for making bricks, they were sent away by villagers who thought it should have been their work. Fathy remarks that sand is 'not a particularly rare mineral in Egypt' (Fathy, 1969: 157), but perhaps this simply emphasizes his distance from the villagers, for whom it was not a matter of the rarity of the material but the etiquette of clan rivalries, a social rather than material concern.

Difficulties increased in the second season of work in 1946, with a plot to flood the site by breaking the dykes which protected it from the Nile's annual inundation, while bureaucrats in Cairo were increasingly obstructive. In 1947, the accounts section of the Department of Antiquities told Fathy he had run out of money, causing work to stop for several months. Fathy records incidents of pressure on local suppliers to overcharge for materials (Fathy, 1969: 163–5), but also that he was summoned to Cairo because the king had taken an interest in his work. It was then, while he was away, that the dyke was cut and the site flooded, destroying some months' work. Then, when Fathy accused the Department for Antiquities of intentionally slowing the project, 'This apparently touched them in a tender spot, for they contrived a clever scheme for getting rid of the whole Gourna project for good' (Fathy, 1969: 161). It was saved only by the refusal of a senior official to countersign a proposal for its closure. But if Fathy realized the bureaucracy was against him because his methods offered no incentives in the handling of contracts, he seems to have been unable to understand the resentment of villagers towards a project he saw as for their benefit – 'they were directly hurting themselves, for they were all earning good money as labourers in the village, and the new houses were, even financially, better than their old ones which, for the most part built on government land, were virtually worthless' (Fathy, 1969: 176).

Fathy had hoped to persuade the Gournis to move voluntarily. Through conversations with village elders he researched the social structures of the old village as a basis for the new plan. He writes: 'we should really have subjected the village to a thorough socio-ethnographic and economic investigation', but he lacked the resources to do this (Fathy, 1969: 53). From his research he determined the spatial organization of Gourna-al-Jadida to be at two levels, that of the family and that of the *badana*, or group of families: 'The *badana* is a tightly related knot of people, consisting of some ten or twenty families . . . [who] live in adjoining houses and, though there are differences of wealth and status . . . follow a communal way of life' (Fathy, 1969: 58). Each *badana* had a cafe, a barber and a grocer; and all families in a *badana* used the same oven when baking bread, and participated in feasts such as circumcisions together. Fathy formalizes this spatiality in New Gourna in a gradation of private to public spaces, with shared courtyards, groups of houses in quarters with small squares, and streets which have detours to discourage strangers. The houses are also grouped irregularly, each having a slightly different plan. Fathy writes: 'by compelling myself to fit the houses . . . into a variety of irregular plots, and by being ready to vary the plan of each to suit the people who would live in it, I made sure that I should think carefully about the design of each one' (Fathy, 1969: 73). Within this, each has a domed space with alcoves for a bed on one side and a cupboard on the other, and a baking oven sited outside the kitchen, which leads onto a courtyard.

The domes are in mud-brick, a technique used traditionally in Aswan and surrounding villages in Nubia, which Fathy first saw on a visit in 1941. He writes of his sight of such domes: 'I was looking at the living survivor of traditional Egyptian architecture, at a way of building that was a natural growth in the landscape . . . a vision of architecture before the Fall: before money, industry, greed, and snobbery' (Fathy, 1969: 7). Fathy found masons in Aswan to bring to Gourna, importing an architectural technique used in Pharaonic times – in the granaries of the Ramesseum (XIXth dynasty), where some mud-brick vaults survive intact today. His justification is that by using mud-brick vaulting the cost of timber used in flat roofs is saved, large durable timber being scarce in this narrow agricultural margin between the Nile and the desert, while local palm trunks would need to be replaced every few years. This is not the only difference between the two villages. The old village is linear, its five clusters of buildings spread along a road, while New Gourna is centred on an open space in front of the mosque. The mosque, too, is of a type found in Nubia, with domed prayer-space, inner courtyard, and two external courtyards. The inner courtyard is planted with trees and shrubs to cool it (Figure 3), allowing cooler air to pass from it, through open-work screens, to the prayer hall. These screens – claustra – are both functional and aesthetic, then, and relate to the wooden *mashrabiya* of houses in old Cairo (Steele, 1997: 39). The minaret has an external stairway, another Nubian motif.

Fathy placed great significance on the courtyard both in the mosque and in houses, or as shared space between houses in which he anticipated members of a *badana*

Mud-brick Utopias 141

Figure 3 New Gourna, the mosque courtyard

would gather, though in Gourna al-Jadida the larger houses have only compounds for animals. For Steele, Fathy's eclecticism is 'defensible on environmental and sanitary grounds . . . [but] unrealistic' (Steele, 1988: 75). On visits to New Gourna I have seen courtyards used for keeping animals but not as spaces for gathering, which takes place on benches in the street so that the street is an extension of the front rooms of houses. Fathy writes:

> In . . . a courtyard, there is a certain quality that can be distinctly felt, and that carries a local signature as clearly as does a particular curve. This felt space is in fact a fundamental component of architecture, and if a space has not the true feeling, no subsequent decoration will be able to naturalize it into the desired tradition.
>
> (Fathy, 1969: 55)

And this may apply to the courtyards of villas and rest houses, though the quote is from Fathy's book on New Gourna. Possibly a more significant factor is the gradation from domestic to public spaces, which Darl Rastorfer summarizes:

> there is an architectural hierarchy to the village which is easiest read by the system of open spaces. The main route to the village interior widens to create a kind of public square around which many of the community functions were to take place including prayer, shopping and entertainment housed by the mosque, *khan* and theatre respectively. The housing is planned in irregular allotments . . . shaping a network of angular streets that turn on themselves to create broken vistas . . . the lives of households are played out in these small, quiet streets that serve as extension to the home and workplace, and as play areas for children.
>
> (Richards *et al.*, 1985: 87)

While the government wanted the Gournis contained, Fathy sought to offer them spaces for sociation, and sought to create the conditions for an alternative economy to that of selling artefacts by introducing village crafts – in alignment, again, with folk-realism but with previously established craft industries. One of the public buildings in New Gourna, opposite the mosque, is the *khan*, designed as an indoor market for textiles and ceramics with rooms for travellers and visiting teachers. A third large, central public building is the theatre, intended to revive folk drama and stick-dancing. Fathy also designed schools for boys and girls in the context of a government drive to build 4,000 schools in the late 1940s, and provided for a market square with trees to provided shade for animals, intended to establish Gourna as a local trading site.

In many ways Gourna is a model village comparable to those, like Bourneville and Port Sunlight in England, that were designed for philanthropic industrialists in the late nineteenth century. The *khan* and the theatre are specifically aligned to national culture, but providing decent housing for the poor articulates a universal value of improvement. Perhaps that is the flaw, inasmuch as the poor's conditions are improved for them, but they are not empowered to make such changes for themselves. But how could they?

New Gourna today

Fathy abandoned New Gourna in 1949, blaming official disdain as the project was shunted around three different government departments, and the few officials who had been sympathetic were moved to new posts. He remarks:

> In all three departments committees sat, apparently convened solely to find excuses for stopping the work and enabling the department concerned to wash its hands of Gourna altogether . . . It was clearly impossible to go on working with such people, so when I was finally told either to return to the School of Fine Arts or to give up my chair and go permanently into the Housing Department, I returned to teaching with relief.
>
> (Fathy, 1969: 183)

Today, after four or five families moved there in the late 1940s and houses were later occupied by migrants from other parts of Egypt (the details of their origins not readily told to outsiders), the village has become a viable community, its economy for the most part in agriculture and local trades, such as cycle repair, rather than tourism. There are no overt signs of abjection now, though the villagers are by no means wealthy. Electricity has been installed, from the High Dam, and several houses have satellite dishes. There is an internet café among the shops on the main road that skirts the village. A few taxis bring foreigners to see what locals have come to realize is a site of interest to them, even if most stay only long enough to be shown the mosque and theatre, not walking through the streets with their blind corners where they would usually be offered tea (Figure 4).

Some houses appear more or less unchanged. In others serious cracks have appeared, caused by changes in the water table, rising salts in the soil and perhaps the extension of adjacent properties. Restoration appears selective as well as sporadic, in a continuation of government indifference mixed with external efforts to ensure preservation. The mosque is cared for by its guardian, one of the original inhabitants of New Gourna, and has been partly restored (supported by donors from other Moslem countries). The theatre has been restored with external aid, too. The school for boys was destroyed and replaced by a large concrete building. The market contains heaps of scrap and is used for unloading lorries, though oxen browse in the shade of trees on the edge of the village. The *khan* is used as a store. Fathy's rest house and field office have been restored and serve as both the owner's home and a small museum. Some houses have been extended, and a few have balconies and tiled facades in imitation of middle-class apartments in Luxor. In 2005, a house owner showed me the additional floor he was building now that his son was married and needed an apartment of his own. It was not yet roofed, as he waited for more money, and built in fired brick and concrete. There were two bedrooms, a kitchen, a bathroom and a roof terrace (Figure 5). Some of the detailing reflected the urban styles Fathy sought to exclude but represented an available vocabulary.

Figure 4 New Gourna, street scene

New Gourna is not, then, preserved like an archaic monument. But it is a living village and not an archaeological site. After more than 60 years, now fully occupied, it demonstrates the viability of building housing for villagers in mud-brick. Where houses have been adapted it has been largely in relation to changes in circumstances, and it may be precisely the capacity to absorb such adaptations that constitutes the viability of the project. Fathy regarded New Gourna as a failure (Fathy, 1969: 3), and so it was in terms of the initial aim of persuading the Gournis to move. But the village as it is today raises the question as to what is an appropriate role for a development architect or planner working in a village setting. Fathy sets aside his own taste when he writes: 'When the architect is presented with a clear tradition to work in, as in a village built by peasants, then he has no right to break this tradition with his own personal whims' (Fathy, 1969: 26), but does not say that villagers themselves cannot extend tradition (though he might have thought that). Now each house has its own supply of running water – while Fathy excluded it as a way to retain the importance of the well as a gathering place for women – but it is difficult to argue against such amenities, just as it is inappropriate to regard the new village as an open-air museum of vernacular architecture. If, then, the project shows the efficacy of a free building material used in a renewed if contrived vernacularism, the difficulty is that the social architectures of community participation were not in place. Yet they cannot be engineered, in Egypt, or in London, New York, Glasgow or Baltimore, where the social housing projects planned and designed in the post-war years to improve the conditions of the poor have since been classified as slums, or destroyed.

Figure 5 New Gourna, recent house extension, exterior

New Baris

After New Gourna, Fathy worked on a ceramics factory and a primary school, and the rural estate of Lu'Luat al-Sahara (Pearl of the Desert). But he left Egypt in 1956 for the architects Doxiades Associates in Athens, working on housing schemes for Arab states and researching humane architecture. He returned to Egypt in 1962 to design a training centre and cooperative for desert development at Kharga (1962) in the New Valley. An hour south of Kharga is the site of New Baris, planned in detail as a 200-acre settlement for 250 families, as the hub of six satellite villages. Nasr saw the New Valley – a name that does not correspond with the area's geography but was used to imply a parallel with the Nile Valley – as a site for development to alleviate the pressure of population growth in Cairo. Underground water allows irrigation and although the area is desert there are oases, as well as some traditional mud-brick villages of uncertain date where flocks of goats roam the streets; small wooden doors lead into the courtyards of houses, the walls of which are thick enough to maintain a more or less even temperature through the year.

Today the town of Baris has begun to expand through renewed government efforts to develop the New Valley. There are fields of barley, rows of shade trees and patches of bright green onions, as well as a few cows. New Baris, a short distance away but deserted apart from its guardian and his family, was abandoned in 1967 at the Israeli invasion of Sinai. All that was built was a market complex for an agricultural cooperative (*suq*), two villas, a house now occupied by the guardian, a bus terminus waiting room, and a car repair workshop, placed at the perimeter of the site to minimize the noise from metal bashing. An administrative centre next to the *suq* is built and vaulted but unsurfaced, the rough brickwork still showing the vaulting technique derived from Nubian examples. In front of it is a pile of unused bricks. Only foreigners visit (currently with a police escort throughout the journey), and in such a dry climate the site is well preserved (Figure 6).

Since the inhabitants were not identified and might come from any part of the Nile Valley, Fathy could not undertake the social research he attempted at Gourna. He had, however, made a tour of vernacular settlements in North Africa, and given a paper on New Gourna to a conference in Cairo in 1960 (Berger, 1964). From this, he devised an orthogonal street plan to maximize shade in summer temperatures of up to 45°C. For Steele, the *suq* is 'the ultimate test of Fathy's attempt to ameliorate the extremely harsh conditions without mechanical means' (Steele, 1997: 141). Only the mosque was to be angled obliquely to the grid, opposite a palm-lined square. Houses have paired courtyards with latticed screens, so that air from a sunlit area rises and draws in air from a cooler adjacent courtyard by natural convection. Breezes are captured at roof level by wind-catches, and filtered down through the house. Thick mud-brick walls act as insulation, as in traditional dwellings. The roofs of the two villas are domed, with decorative lattice-work screens to the roof terraces (Figure 7). Wind-catches are used to great effect in the agricultural cooperative where the temperature in storage vaults below ground level – designed for storage of produce brought to market in the shops that line the central courtyard (where there are also baking ovens) – is between 15 and 18°C in summer. On my visit in February, 2005, the outside temperature was 25°C and the vaults

Figure 6 New Baris, the agricultural cooperative

were cold. The *suq* has an iconic quality, its vaults and screens in strong light and shadow in the uncanny silence of the deserted site. I found it almost hypnotic. But I was a visitor, and it is more significant that Fathy designed the *suq* for a cooperative. He writes in a paper to the Egyptian Society of Architects (1977):

> the only alternative is to resort to the traditional cooperative system by finding means to make it work under the non-traditional conditions prevailing nowadays. The snag in cooperative building is that one man cannot build a house, but ten men can build ten houses easily.
> ('Bariz Case Study', in Richards *et al.*, 1985: 92)

This paper was a response to a UNESCO report on housing in developing countries, the area Fathy had by the 1960s made his preoccupation. Fathy realized at New Baris that the traditional skills of vaulting in mud-brick on which he had relied at New Gourna did not exist in the region and would need to be taught, affirming a role for the architect as intermediary. He writes: 'these traditions do not exist any more in most peasant societies and it is implicit that we secure the assistance of the specialized architects to revive the lost experience and traditions among the peasants until a new tradition is established' (ibid.). For this to function, he adds, the families need to know each other, and the critical mass needs to be small enough – around 20 families – to avoid anonymity in the collective. The question is whether such an attitude can be introduced into a situation through professional agency or must arise at grassroots level. The abandonment of the project meant this idea was

148 *Horizons*

not tested, but it is common in societies where mud-brick is used for people to cooperate in the regular resurfacing of walls that ordinary weathering requires. It does not seem fanciful to organize cooperation in the building of a settlement as well, as happens in the affluent world in self-build schemes. Fathy also argues that the architect can design individually varied dwellings within a large scheme:

> when the architect is faced with the job of designing a thousand houses at one time, rather than dream for the thousand whom he must shelter, he designs one house and puts three zeros to its right . . . But the architect has at his command the prosaic stuff of dreams. He can consider the family size, the wealth, the social status, the profession, the climate, and at last, the hopes and aspirations of those he shall house . . . let him begin with the comprehensible, with a handful of people or a natural group of families . . . Once he is dealing with a manageable group . . . then the desired variety will naturally and logically follow in the housing.
>
> ('Bariz Case Study', in Richards *et al.*, 1985: 93)

This is not fanciful, either, and can be observed in settlements such as the villages of the Dra Valley in Morocco (Dethier, 1982: 18–19), maintained through collaborative efforts between families. Of course, there are no architects to design such traditional settlements.

A housing revolution?

Steele asks why Fathy's work did not ignite a housing revolution for Egypt's rural poor (Steele, 1988: 74) but the answer is obvious. For the villagers of Gourna the new village was not about better living conditions but about the authorities' wish to deprive them of their traditional livelihood. Additionally, the rural poor had no power or resource to ignite change, nor any viable precedents from which to derive a vocabulary of possibilities on which to base such a revolution, while the political revolution of 1952 led to other kinds of changes more immediate to their material lives, such as land redistribution and universal education. Had New Baris been completed, it would have housed families from different places with no guarantee that they would choose to associate across extended family lines. In design terms it deals with the area's climate through the no-cost, natural energy means of wind-catches and convection, on which Fathy wrote a second book (1986) in which he illustrates part of the plan of New Baris together with an aerial photograph of Marrakesh showing its predominance of courtyards. This aspect of Fathy's work is of interest today in the context of natural building methods. But again, a difficulty is that social architectures – cooperative building methods, for instance – are overlooked when attention focuses on design and the aesthetic appeal of Fathy's work in the West. This leads to a different line of argument: Darl Rastorfer, like Steele, asks why Fathy's solution to the problem of providing rural housing remained marginal:

Figure 7 New Baris, villa, detail of roof terrace

Could there be a critical element missing in his framework that prevented his social-oriented projects from being widely replicated? Fathy had developed his architecture of private houses by seeing many designs realized. Had he been given more public support for community projects, then too, he may have refined or evolved a self-sustaining solution. Others see the basic reason his projects have not been replicated as institutional: land tenure, public services and utilities, and finance.

(Richards *et al.*, 1985: 94)

Fathy was obstructed by government bureaucrats but Rastorfer hints that the underlying obstacle was a failure to facilitate social processes, thereby limiting the scope for further application in grassroots projects. As Rastorfer says:

New Gourna's identity is related to similar villages in the region, but it possesses a character all its own and if there are shortcomings they are measured in the gap between intention and realization. The planning intentions far exceeded the rural population's ability to upgrade itself. Though community life in rural Egypt is undergoing transition, people have not yet broken fundamental ties.

(Richards *et al.*, 1985: 89)

Fathy sees the peasants as unable to conceptualize space:

The Gournis could scarcely discuss the buildings with us. They were not able to put into words even their material requirements in housing; so they were quite incapable of talking about the style or beauty of a house. A peasant never talked about art, he makes it.

(Fathy, 1969: 40)

This is a not untypical urban-professional view. I would add, then, to Rastorfer's point (cited above) another made by Suha Özkan, writing that Fathy challenged imported material and technologies, while his architecture:

though directed at the poor, had the support of many of the upper classes who supported the development of an indigenous culture. In order to display the validity of his architecture, Fathy did not hesitate to build for the rich. However, his discourse and architecture could not achieve their intended result of harmonious living imbued with cultural values and historical continuity. This was because they were based on traditional methods and modes of building at the very time when the Egyptian communities he was building for were undergoing radical transformation. Fathy's technical solutions were rendered irrelevant by changes in societal values.

(Özkan, 1997: 41)

Özkan goes on to say that Fathy's thinking remains relevant despite the failed application of his ideas. Fathy admitted his social research was limited, and his account of New Gourna implies a romantic idea of the countryside. The clan structure on which his spatial order was based does not operate in New Gourna because most of its inhabitants come from elsewhere. But this does not mean the villagers do not have a common culture, even if it resembles the culture of common interest characteristic of urban settlements more than the family or land ties of rural areas. I recall that most houses I went into in New Gourna had framed photographs of President Nasr in the front room, often next to photographs of family members, and that many men in New Gourna had worked on the construction of the Aswan High Dam (a project of Nasr, which Fathy vehemently opposed). Such ties of common interest add weight to Rastorfer's argument (above) that Egypt was changing at the time Fathy sought to re-inflect its traditions.

An other Modernism?

Rastorfer's suggestion of a modernizing side to Fathy's work invites comparison with Western Modernism. J. M. Richards writes that Fathy's ideals coincide with those of Modernism 'to a degree that he might find difficult to accept' (Richards *et al.*: 10). Steele says this remark 'caused a storm of protest among followers by suggesting that his functionalism . . . put Fathy on a parallel path with Modernism' (Steele, 1997: 183). He claims Fathy disputed this, but goes on to say that Richards does not go far enough. Richards, a participant in the International Congress of Modern Architecture (CIAM), saw Modernism as aiming to deliver better living conditions for a range of social classes, a utopian project (Curtis, 2000: 56), and makes much of Fathy's efforts to link architect and dweller. His critique is of an 'utter unfamiliarity . . . [that] wounded people's instinctive desire to feel at home in their surroundings', compounded by a privileging of professional expertise that 'led to an urge . . . to destroy the past in order to create a better-ordered future' in Modernism (Richards *et al.*, 1985: 11). Fathy, in contrast, represents a process in which the architect relates to the dweller – a prescription for a redemption of Modernism. Richards continues:

> [Fathy's legacy] includes the message that in solving human problems one must not remove oneself too far from the human individual, and that even, perhaps especially, in our age of infinite technical resources, the simplicities inherent in the very nature of building must not be overlaid by the worship of progress.
>
> (Richards *et al.*, 1985: 14)

Richards remarks of CIAM's 1951 meeting that 'the world of the architect had suddenly expanded to embrace that of the town planner and even the sociologist' (Curtis, 2000: 56), and argues that Fathy's reconstitution of a vernacular is adaptable:

> There are . . . in the spaces between the buildings, proportional variations infinite in number, invisible maybe to the observer but giving satisfactions to

the user of which he [*sic*] may not even be conscious. Fathy's use of mud-brick moreover is not an archaic fad, a choice of primitive methods for a primitive way of life. It can be admired in terms of the strict scientific use of available materials, of the relation of building costs to habitable space, and of optimum thermal efficiency.

(Richards *et al*., 1985: 14)

When he is critical of architects who import design, Fathy is arguing, I suggest, against a corruption of rural life by economic colonialism. Perhaps, too, an imported vocabulary does not only debase indigenous culture but is also a way in which a colonial power assimilates the elite of the subject state (as Fanon argues). Fathy writes:

the work of an architect who designs, say, an apartment house in the poor quarters of Cairo for some stingy speculator, in which he incorporates various features of modern design copied from fashionable European work, will filter down, over a period of years, through the cheap suburbs and into the village, where it will slowly poison the genuine tradition.

(Fathy, 1969: 21)

But he adds that 'a tradition . . . may have begun quite recently. As soon as a workman meets a new problem and decides how to overcome it, the first step has been taken in the establishment of a tradition' (Fathy, 1989: 24). The determination of tradition, then, like the specifics of design, takes place in the material culture of building, hence in decisions that are taken, as in craftwork, in the process of manufacture, and lead to constant variation within a set vocabulary and material means. This collapses the dualism of designing and building as conceptual and material practices which is the convetion of Western modernity, or, in Henri Lefebvre's terms (1991), redirects emphasis from the conceived space of plans to the lived spaces of dwelling. Fathy saw this in utopian terms. Like Western Modernists, he interpreted the problems of the poor but did not hand over the power of interpretation to them. Yet his work remains progressive in its cooperative vision and the appropriateness of the technology of a no-cost architecture; and perhaps this has relevance now to the housing needs of affluent countries when climate change forces a search for low-impact technologies of dwelling.

Doshi and enablement

Finally, I want to cite the work of contemporary Indian architect Balkrishna Doshi, and recent housing projects in the non-affluent world. Steele says that Doshi's nationalism and desire for autonomy equal Fathy's, though 'His emphasis has not been on deriving a clearly recognizable "Indian architecture", but on a logical, theoretical archaeology that will lead him to the principles behind the form' (Steele, 1998: 90). But how does Doshi address social architecture, and how does this compare to new projects guided by radical thinking in non-governmental agencies?

Doshi's house and studio at Ahmedabad, the Sangath (1979–81) – meaning 'moving together through participation' in Sanskrit (Steele, 1998: 85) – has a vault based on the Platonic circle, a semicircle visible above the roofline that if completed would touch the floor. At a distance, the mix of rectilinear and curved forms resembles the style of New Baris, but on closer inspection, apart from the use of concrete and ceramic surfaces, the feeling is more urban and industrial, and less vernacular. Similar principles are applied in the Gandhi Labour Institute at Ahmedabad (1980–4). Steele writes of the latter, perhaps a little romantically, 'The individual vaults each represent a house cluster around the village square, whose green replaces the farmer's fields' (Steele, 1998: 112). The vault is made from clay plugs manually covered with concrete into which pieces of broken china have been set, and was intended as a prototype to provide an 'inexpensive, remarkably strong yet light, and environmentally intelligent' solution for a hot climate, the clay contributing high insulation from, and the white ceramic surface high reflection of, the heat (ibid.). But perhaps the most relevant of Doshi's projects for this chapter is his housing scheme at Aranya, Indore (1983–6).

India is a rural country, around 70 per cent of its population living in villages. But urban populations have increased with industrialization and in enclaves within the large cities, or on their peripheries informal settlements have multiplied. Periodically, as land prices rise and their sites are taken for development, they are cleared. Many poor people live on city streets in makeshift shelters that do not constitute even informal housing, while the government housing schemes provided to alleviate their conditions are ineffective, built in inappropriate materials on sites that no one else wants, and inflexible. In contrast, in India, Pakistan, Sri Lanka, Jordan and South Africa, among other countries, a current approach is to provide plots with services such as energy, clean water and separation of sewage from the water supply, and then hand over the means of building to local people (Hamdi, 1995: 88–108; Umenyilora, 2000). At East Wahdat, Jordan, in the context of an urban population one-quarter of whom live in informal settlements, an upgrading project in the late 1980s offered basic service units and a legalization of settlements. As Ismaïl Serageldin writes, 'rather than raze the shanty town, they empowered the people to transform it' (Serageldin, 1997: 94). People were offered purchase of plots, with credit, and were able to move their existing shelters to a corner of the site while construction took place, dwellers completing building work after the service unit and first room had been installed. Serageldin continues:

> The result is a thriving community with plots of varying sizes (ranging from 60 to 200 square metres) according to the needs and means of each of the 523 families. All plots are provided with water, sewerage, electricity and road access. Schools, clinics, community and vocational training centres were built and technical assistance was made available to the families to help them to build their own houses as well as fifty-eight shops and twenty-four workshops.
> (Serageldin, 1997: 94)

A similar if less radical approach was used, at around the same time, in the Orangi Pilot Project Housing Programme in Karachi, Pakistan (Hasan, 1997), working through local building suppliers and contractors to adapt methods and provide training for masons, and aiming to provide a basic set of components (though in concrete block form) for self-builders to legally take over.

A site-and-services approach of this kind was also used at Aranya, where building materials could be purchased from a local cooperative and the cost repaid over time. Steele records that Doshi found, studying local informal housing, that there was 'a legible pattern, with huts clustered into small neighbourhoods and houses with public zones opening onto common spaces'. There were shops in all cases, including the most deprived, and tree planting in common spaces (Steele, 1998: 116). Streets are used as both through routes and extensions of domestic space – as in inner-city neighbourhoods in the affluent world. The development is mixed-income, though was planned mainly for the poor, initially housing 6,500 people on an 86-hectare site bounded by the Mumbai–Delhi highway on one side. Research carried out by the Vastu-Shilpa Foundation for Studies and Research in Environmental Design led to a master plan. Lailun Ekram lists its objectives (Ekram, 1997: 106–8):

- to create a new township that has a sense of continuity with existing settlement patterns, retaining the fundamental values of security, and to plan a good living environment;
- to achieve a settlement/township character by harmoniously integrating people and their built environment;
- to create a balanced community of the various socio-economic groups, encouraging co-operation, fraternity, tolerance and self-help generated through a physical planning process;
- to evolve a framework where incremental development can take place within a legal, economical and organizational framework.

The land is divided into plots of 11 sizes, from 35 square metres (for the poorest dwellers) upwards, and options were given for the extent of services and features such as porches, balconies and open stairways. Houses are clustered around shared courtyards opening onto the street (in an architecture designed for, if not actually producing, social uses of space, as in Fathy's work). But plots were subsequently traded by those to whom they were allotted, so that prices increased by up to 10 times in speculative transactions. Steele comments:

> There are a number of critical multi-disciplinary inputs that are missing at Aranya. First, there is no evidence of any activity by non-governmental organizations concerned with community participation. Secondly, no identifiable arrangements for credit and mortgage had been put in place to enable slum [sic] dwellers to secure a plot. Thirdly, no arrangements for providing gainful employment had been made.
> (Steele, 1998: 122)

Hence housing for poor people which they could build themselves has in part become a middle-class enclave. But there is also a tacit hierarchy within the spatial organization of Aranya, the smallest (poorest) plots being in the centre of the development and the plots for middle-class buyers (14 per cent of the total) on the edge by the road. In one way, then, the success of Aranya demonstrated in its uptake by middle-class dwellers contradicts one of its premises, to provide integrated housing for the poor, which raises issues as to how a project can effectively be handed over, and the extent to which its future is determined by forces within the mainstream society (including market forces). The answer may be to strengthen regulation, though this needs to be compatible with handing over power.

To take another case, township dwellers in South Africa generally ignored, or took and rented out, the government housing provided to relocate them from the townships as a means of controlling them under the white regime; but they had experience of building informally in the townships (which were periodically razed, then rebuilt). They have used this, with a basic service provision model, to create flexible housing in a standard of construction and materials higher than in illegal settlements. Chinedu Umenyilora writes that the outcome is that 'exquisite and radical forms of architecture spring up from within communities' (Umenyilora, 2000: 213). It may be that an internal market operates in such developments, too, but my impression – from published accounts such as Umenyilora's, and stories of the rejection of privatized water companies who raise charges for drinkable water by not paying the bills – is that there is a sense of grassroots equity. The lesson seems to be that without appropriate social architectures – which are constructed by the mobilized community in South Africa – schemes are unlikely to be sustainable despite their quality of design, the adaptation of a vernacular style or use of locally available materials. In the context cited here, social architectures include provision of economic assistance and for acquisition of skills, as well as a basis for collaboration. Nabeel Hamdi writes, 'Participation without enablement is like trying to drive a car without fuel', while enablement is not an abstract idea derived from 'social science, public management, political radicalism, benevolence, or philanthropy' but a materially articulated process (Hamdi, 1995: 88).

To conclude, the logical extension of Fathy's enlightened efforts to create a no-cost architecture for a non-affluent country may be the move to basic service provision and a handing over of design and building, as well as control, to dwellers. The evidence from many projects in non-affluent countries today is that this is viable, requiring a minimum of investment in energy, water and waste treatment, together with a legalization of self-build housing. No one calls any such cases ecovillages, which they are not; nor does the practical everyday grassroots organizing that goes on in self-build, or in locally managed, participatory housing schemes, constitute activism. And yet, in the terms of Chapter Seven, the glimpse of liberation enabled by activism or ecovillage living may be a Western concept, and a different concept will be necessary if the term is to be mapped onto post-colonial situations. If Fathy's work is flawed by the same privileging of the role of the architect as interpreter of reality that characterizes Modernism in the West – though Fathy used his professional status to facilitate projects, and his approach was grounded in an emergent national cultural identity, a residual vernacular tradition and a refusal of the West – the effect is not, I would argue, that his work should be dismissed, but that extension of the model to include a handing over of planning to grassroots groups is entirely compatible with the fusion of design and building that Fathy's limited recasting of the architect's role achieved, in a sustainable material culture.

Despite the extent to which such a shift occurs in development settlement planning today (Serageldin, 1997), the literature of ecovillage living ignores such evidence except as a site of spiritual traditions. Hildur Jackson and Karen Svensson, for example, as editors of *Ecovillage Living* (2002), cite few non-Western cases in sections dealing with ecological and social issues, and look to the non-affluent world only under the heading of the cultural-spiritual dimension. I wonder whether sustainable settlements are as open as this implies to reinterpretation within the framework of a spiritual life supposedly lost in the industrialized West, which never was, or if this rehearses the Orientalism of the late eighteenth century.

9   A barefoot society

Introduction

In the previous chapter I examined Hassan Fathy's mud-brick architecture, finding that while Fathy's utopian vision is relevant to debates on sustainability today, it retains a reliance on professional expertise, and lacks the social architectures by which dwellers shape habitats for themselves. In this chapter, I focus on the work of the Social Work Research Centre (SWRC), or Barefoot College, a rural campus in Rajasthan, India, built by villagers with no architectural training. The Barefoot College was founded by a group of urban professionals in 1972 to work on water harvesting, but is now a multi-purpose campus for villagers' education based on Gandhian principles of non-violence, voluntary simplicity, and working with the poorest of the poor so that they may thereby help themselves.

I begin by outlining the development of the Barefoot College, its basis in Gandhian principles, and its programmes. I ask how these contribute to new individual and social awareness in the context of post-colonial discourse and a pedagogy of liberation, not claiming that the Barefoot College is Utopia but asking how insights gained from its experience might be applied in the affluent world. Finally, asking what it could mean to work with the poorest of the poor in the West, I note the work of the Coventry Peace House, a commune working with asylum seekers in the English West Midlands.

SWRC and the Barefoot approach

The Barefoot College began in 1972 when Bunker Roy (its present Director) and a small number of colleagues arrived from Delhi to work on water harvesting, using the British-built Fever Hospital at Tilonia as their makeshift base (Figure 8). The group sought to work directly with villagers, outside the professional bias of the Indian social work tradition. The hospital had been used for storage but provided a rudimentary base in a main administrative building, adjacent wings now used as workshops, and a compound.

The first problem addressed by SWRC was to provide adequate drinking water in the vicinity, and the second the conservation of water for irrigation. Tilonia is on a plain surrounded by hills, with an annual monsoon. However, deforestation in the Himalayas has caused changes in the weather patterns that produce the

Figure 8 SWRC, the old Fever Hospital

monsoon, diminishing the level of rainfall and leading since the mid-1990s to an earlier onset of the hot season. The monsoon has failed in five years since 1995, causing losses of cattle and crops to local farmers. The problem of water is therefore urgent if the area is not to slide from marginal agricultural use to desert, while erosion following loss of trees cut for fuel for cooking adds to a need to find alternative energy sources as well as to harvest water. This was perceived by SWRC as more than a technical question, requiring efforts to empower as well as educate local communities, to make the most of local knowledges and skills, and to involve villagers fully in the development of new approaches to the social as well as natural environment.

In 1986, the Barefoot College acquired a new site, while retaining the Fever Hospital. A ground plan for the new campus was drawn by an architect in Delhi but no drawings for individual buildings were made. The plan was marked out on the ground with sticks. Building began using local stone, villagers' tacit knowledges and the labour of staff and villagers. After four years the 80,000-square foot (7,400 m^2) site provided residential accommodation for staff and villagers undertaking training courses, with a 10-bed referral hospital, pathology laboratory, pharmacy, water-testing laboratory, and spaces for teacher training, solar technologies, a puppet workshop, a video-editing and screening suite, a screen-printing workshop, a library and an open-air covered stage. Craft workshops, including facilities for cotton weaving, remain in the Fever Hospital, with a metal workshop. As well as the campus more than 200 houses were built for homeless people in the surrounding area, all by barefoot architects. The new campus can house up to 200 villagers participating in residential training, with 50 staff who live permanently on site. There is a guest house, common dining room, meeting hall, post office, administrative centre, internet hut and telephone hut. A craft shop sells goods produced by villagers at SWRC, and there is a 700,000 litre rainwater harvesting tank. The campus is lit by solar energy from roof-mounted panels, and a solar cooker on the kitchen roof provides all cooking except for chapatis – made on a small hot plate fuelled by bottled gas. Grey water is recycled for irrigation. A decision was taken not to be self-sufficient in food, however, because buying in local markets offers a way to relate with local communities and farmers.

The Barefoot College gained an Aga Khan Award for Architecture in 2000, at which point the architect who drew the plan sought credit as its designer. Roy responded that it is an established practice in modern society to ignore the contributions of ordinary people who 'have thus remained invisible . . . in spite of their brilliant creativity . . . because they were poor and illiterate' (www.barefoot college.org). These contributions are recognized by the Aga Khan Award panel, which made awards to projects including a rural self-help and water harvesting project in Morocco, a poultry farming school in Guinea, a village for orphans in Jordan built in a vernacular style and a social centre in Turkey, as well as to the Barefoot College (Aga Khan Award for Architecture, 2001). This is in contrast to the more frequent marginalization of the efforts of dwellers in informal urban settlements in the non-affluent South to build social stability, and to build settlements with materials categorized as waste. The Aga Khan Award literature, however, states:

> The barefoot philosophy is based on a belief that . . . village communities thrived with no paper-qualified doctors, engineers, architects or teachers. Such communities developed their own knowledge, which has since been devalued. . . . The college aims to demonstrate that village knowledge, skills and practical wisdom can be used to improve people's lives – an attitude of self-help drawn from Mahatma Gandhi's example. Since rural people already have tried and tested traditions, the college believes that all that is needed is a little training and upgrading, combined with respect for local skills.
> (Aga Khan Award for Architecture, 2001: 78)

In the same volume, Kenneth Frampton, citing John Turner's research on informal settlements in Latin America (Turner, 1963; see also Turner 1976) and Hassan Fathy's *Architecture for the Poor* (1973), writes of the limits of architecture in the face of global poverty. On the Barefoot College, he says:

> Based on a design that seems to have been arrived at on a collective basis, the architectural result is at once both surprisingly formal and informal. The basic building syntax itself could hardly be more strict and severe, even though the technology employed is quite hybrid, so that it is not something one could possibly recognize as vernacular in the traditional sense.

> In the last analysis, the Barefoot College displays all the traits of Utopian community ... One is irresistibly reminded to an equal degree of both Rabindranath Tagore's Santiniketan College and of Charles Fourier's paradigm of the *phalansterie*.
>
> (Frampton, 2001: 12)

I would question a resemblance to the phalanstery (Chapter Three) in that the principles on which SWRC operates have little in common with Fourier's tabulations, or classification of people into psychological profiles (as they might now be called), nor with his model of a community selected to match all such types rather than on a basis of need. And if the architectural syntax to which Frampton refers is a product of the concept of voluntary simplicity, this is not a visually impoverished environment, as climbing plants cover the walls of buildings and paths are lined by trees (Figure 9). The hybridity to which Frampton refers may denote the use of plywood, sand and lime as well as stone, pragmatically reflecting the availability of resources and stretching of local skills in making large buildings, but may also relate to the presence of several geodesic domes around the site (and in some villages).

Gandhian principles

The Barefoot College aims to benefit the poorest of the poor 'for whom there are no alternatives', addressing poverty, discrimination, injustice, and exploitation,

Figure 9 SWRC, street with post office

providing a place where the poor can 'be heard with dignity and respect, be trained and be given the tools and the skills to improve their own lives' (www.barefoot college.org). In India, 70 per cent of its population being villagers, and despite the high visibility of poverty in city streets, the rural poor are among the poorest. The college states the following principle:

> The Barefoot College is a place of learning and unlearning. It's a place where the teacher is the learner and the learner is the teacher. It's a place where no degrees and certificates are given because in development there are no experts – only resource persons. It's a place where people are encouraged to make mistakes so that they can learn humility, curiosity, the courage to take risks, to innovate, to improvise and to constantly experiment. It's a place where all are treated as equals and there is no hierarchy.
>
> (www.barefootcollege.org)

This is implemented in consensus decision-making at monthly business meetings open to all staff. Meetings can vary the salaries of individual staff for a fixed period – in one case reducing that of one person to discourage smoking – but are more often concerned with programme development. Salaries are in any case based on the national minimum wage with additions for the responsibility of coordinating an area of work, up to a maximum of 4,500 rupees per month. SWRC thus operates on a small budget and has relatively little need for external income. Such voluntary simplicity is part of the ethos of the campus, in keeping with the commitment of the urban professionals who established it to integrate with village society by living full-time on campus rather than in a nearby town where a more conventional professional lifestyle could be retained. This extends a more general engagement with village communities that began in India among enlightened urban professionals in the late 1960s but did not at that time lead to immersion in rural life or identification of issues from within a rural community, nor to the endorsement of local knowledges that underpins the work of SWRC.

In the 1970s, SWRC worked with four or five villages. Few literate people were able to represent villagers' interests to the authorities, and concerns were identified through a series of village meetings. The process echoed but also widened a traditional rural social architecture of decision-making by village elders: whereas this excluded women, SWRC has given particular attention to enabling village women to gain independence within village societies. Similarly, the caste differences of traditional Indian society, in which the poor suffer social as well as economic exclusion, are addressed in the campus ethos of equality and mutual respect regardless of background. Since completion of the new campus, SWRC has concentrated on residential training courses for villagers, with some programmes led by women, and developing a network of night schools linked to water harvesting and health care. The ethos indicated above is presented, too, in a code of conduct for staff and students at the Barefoot College, which, for the sake of accuracy, I cite in full:

- Live and work in close proximity with the rural community;
- Create a space for creative and constructive personal growth – not discriminating against caste, religion or political thinking;
- Ensure gender equality within the organization;
- Have an intrinsic belief in the democratic political process and not follow partisan agendas or include partisan politicians on the board;
- Judge the worth of people by their willingness and ability to learn – not by their paper qualifications;
- Believe in the law of the land and have a commitment towards social justice through non-violent means;
- Have respect for collective, traditional knowledge, beliefs, wisdom and practices of the community;
- Be committed to the preservation of natural resources and not endorse processes that destroy, exploit or abuse natural resources;
- Use appropriate technologies that sustain the community and not encourage technologies that deprive people of their livelihoods;
- Set a personal example in adhering to the code of conduct.

(www.barefootcollege.org)

Aspects of the code reflect Gandhian non-violence, non-sectarianism and concern for the appropriateness of technologies. Gandhi remains a national icon, his image still shown on all Indian banknotes. I was told his ideas had been formative for the post-independence generation of SWRC's founders, though he is read less by young people today. The issues of technology and industrialization have a particular relevance in the context of current efforts to industrialize India, and it may seem that Gandhi's ideas have become outmoded. Yet Gandhi was not against machinery as such, only that which was detrimental to human well-being and social cohesion. While, then, Gandhi's example of spending an hour each day spinning is sometimes taken as such a rejection, it is more accurate to say he sought industrialization appropriate to the maintenance of village society. William Shirer cites an interview with Gandhi in 1931 in which Shirer says: 'you said your idea of an ideal state would be one without factories, railways, armies or navies and with as few hospitals, doctors and lawyers as possible' (Shirer, 1981: 36–7), a position which Gandhi confirmed in reply. Glyn Richards, however, links Gandhi's rejection of industrial development to the philosophy of simplicity and self-realization, summarizing Gandhi's contention as being that economic policy cannot be separated from moral values. Hence, an economy should provide for everyone to feed and clothe themselves while, 'should anyone seek to possess anything above and beyond' that necessary for her or his immediate needs, 'it could be construed as stolen property and indicative of lack of faith' (Richards, 1991: 113). In the face of the existing wealth of the upper strata, Richards argues that a Gandhian position would be that such wealth is held in trust, allied to a call to adopt bread-labour – spinning, weaving or carpentry – as well as intellectual work. Richards adds that Gandhi was revolutionary here, 'since on the whole in India manual labour was regarded as the duty and prerogative of the lower stratas of society' (Richards, 1991: 117).

This leads to a defence of labour-intensive forms of work but is not a rejection of machinery that can improve the quality of such work. According to Richards, Gandhi looked back to a time when India was self-sufficient in cloth and food, before home spinning declined with the British introduction of textile mills: 'A two-fold loss ensued: the loss of labour . . . and the loss of income required to purchase clothes from the mill' (Richards, 1991: 118). A revival of village industries thereby remedies the situation in restoring self-sufficiency and self-respect, represented in the establishment of the Village Industries Association. Richards then clarifies that this is not a rejection but a critique of industry:

> [Gandhi] . . . maintains that hand-spinning is not meant to displace any existing form of industry . . . Its main purpose is to 'harness every single idle minute . . . for common productive work.' The cloth produced in the mills provided work for a limited number; home-spun cloth would give work to all. . . . 'Khadi [home-spun] serves labour, mill cloth exploits it.' It was because of the impoverishment of millions of Indian peasants resulting from the importation of cloth from England that Gandhi could refer to the practice as immoral. . . . To wear khadi, therefore, was an expression of one's solidarity and kinship with one's neighbours.
>
> (Richards, 1991: 119)

I have made what might seem a detour into Gandhi's view of industrialization because it draws out relevant arguments. First, the purpose of work is practical and a means of self-realization, a socially just organization of work providing for the well-being of all, not a division of rich from poor. Second, technologies are viewed critically for their impact on human well-being. This allows a negotiated acceptance of industry on principles of social equity and the equality of manual and intellectual labour. Third, in the context of Indian independence, certain kinds of industrialization and economic practice are not simply carried over from the colonial era but are inherently immoral.

I want to take this further because it appears to underpin the work of the Barefoot Campus – not an official Gandhian institution but expressing Gandhian values – in its aim to empower villagers. Richards depoliticizes the issue while Vivek Pinto, in discussion of Gandhi's agricultural policies, writes that Gandhi envisaged a decentralized India, each village being 'a little republic, economically self-sufficient, and politically autonomous' (Pinto, 1998: 131). He cites Gandhi's *Constructive Programme* (see Gandhi, *Collected Works*, vol. 75, pp. 146–66), to argue that Gandhi sought to set up village cloth production as 'a non-violent weapon for political non-cooperation and economic boycott of British textiles flooding the Indian market' (Pinto, 1998: 134). Homespun cloth was a political material and a basis for indigenous development parallel to the improvement of cooperative agriculture. Pinto continues that other appropriate village industries include soap-making, hand-grinding and pounding, paper-making and match-making; and that these are a basis for a national cultural-industrial identity, or national taste (Pinto, 1998: 135). Today, when Indian mass-produced synthetic textiles flood the market, developing local industries that 'give employment to villagers and profitably use their local resources for meeting their own needs' is a priority (Pinto, 1998: 134).

Water harvesting and solar power

As stated above, the monsoon has failed five times in recent years, while the monsoon rains are lost in run-off without water harvesting. SWRC has worked to collect and store rainwater, its own water-collecting tank being beneath the theatre and fed by a network of pipes from roofs and other catchments. This supplies a hand-pump in the centre of the campus and taps in buildings such as the kitchen, dining room and guest house. Thirteen nearby villages subscribe 30 rupees per month to maintain connections to harvested water in a system designed and built by villagers themselves. Water from the surrounding hills is channelled using the natural flow and collected in tanks at more than 400 schools and community centres, where it is accessed by villagers and children attending night schools (see p. 110). Water could be drawn from underground sources using pumps, or taken from village ponds, but much of the groundwater in this area is brackish or otherwise unfit for human consumption, pumps are expensive to a village economy, and continuous drawing up of water from wells can reduce the water table. A large number of open village wells have therefore been converted for water storage, while more than 200 ponds have been cleaned of silt as a long-term measure. Mobile water-testing kits are made available, and 45 village young people have trained as barefoot chemists able to take and analyse water samples. Fifteen women have trained as barefoot engineers competent to construct overhead water tanks, and 250 villagers have trained in drought prevention programmes.

Despite these efforts, the region around Tilonia faces a drier future as a consequence of climate change. The use of trees as fuel is therefore a particular threat in that tree loss leads to further environmental deterioration. While water harvesting was initially, and remains, a priority, the programmes for solar lanterns and solar cookers complement it in building renewable energy village-based economies. The campus is self-sufficient in energy (apart from the chapati hot plate), the capacity of its roof-mounted solar panels being 40 kilowatts, enough to supply about 4,000 low-energy light bulbs. On the main campus, villagers, mainly women, from several Indian states and some other non-affluent countries learn how to make and maintain solar lanterns that, if placed in the sun during the day, provide four to five hours of light at night, and are used in every night school. They have been taken to village communities from the Himalayas to Tamil Nadu by women who trained in their manufacture and maintenance at SWRC, and are used in Ladakh, Afghanistan and Ethiopia as members of village communities there have come to the Barefoot College to gain solar engineering skills. A total of 3,530 solar lanterns were manufactured by 2007, and around 400 villagers have now trained as solar engineers. In the Himalayan region the use of solar lanterns and cookers reduces local deforestation in a way that may in time mitigate the impact of climate change on Rajasthan's monsoon, in the context of long-running grassroots campaigns against forest exploitation (see Chapter Six).

Solar cookers require materials which must be bought in, such as the mirror surfaces, and, like solar lanterns, require precision, but the metal framework can be manufactured, given adequate welding skills, from scrap metal from disused agricultural machinery, as it is by villagers at SWRC working (in 2004–5) with a small number of European volunteers in metal workshops in the grounds of the Fever Hospital (Figure 10). The metal framework is covered on the concave side by small mirror plates focusing the sun's rays to a precise point at which the maximum heat is produced. Demonstration of a solar cooker involves holding a rolled newspaper at this point, when it bursts into flames, while a cooking pot mounted at the focal point will absorb the heat to boil whatever is in it. The solar cooker was not invented by SWRC – thousands are in use around the world, as on the roof of the solar kitchen at Auroville, where one large cooker provides hot food for 1,000 people on a daily basis (see Case Three) – but it is key to village self-sufficiency and sustainability, decreasing deforestation and reliance on the other major fuel source, kerosene.

Communications and rights to information

Training in solar engineering requires workshops and equipment and necessarily takes place on campus. Video editing, digital media and website management also have their own facilities at SWRC, while traditional means of communication are employed in outreach activities, taking puppet theatre, street theatre and songs to villages throughout the state of Rajasthan. The aim is to provoke dialogue, and plays tend to revolve around a mix of everyday themes such as water, health and social issues such as the role of women or village relations with the authorities. The scripts are improvised and audience members are encouraged to become involved. For example, in one street theatre play, an inspector visits a village to decide where a well should be dug, ignoring the advice of villagers and elders and insisting on following his own, urban expertise. He commands the work but when the well is dug it is dry. The villagers dig another well, finding water, leading to general rejoicing. A leaflet produced (in English) by the communications team at SWRC states that it seeks to provoke dialogue on 'social, economic, political, cultural and environmental issues', including caste discrimination, health, education, collective decision-making, access to information, and human rights. Just as wearing homespun cloth was a political act for Gandhi, popular culture has a political dimension articulated by a performative means and linked to SWRC's concern for villagers' rights to information. The tradition of puppet theatre has been adapted by the Barefoot College to transform it from a high to a popular art form. While the tradition of puppet-making in Rajasthan is based on carved wood string-puppets, and characters tend to be kings and queens, princes and princesses, nobles and other actors representing an elite, playing out epic tales remote from the lives of villagers, the Barefoot College puppet workshop has evolved its own tradition over more than 20 years, based on glove puppets made from pulped paper, using newspapers and old reports. These are brightly painted, and some are based on staff and visitors, or villagers from the audience. Puppet plays are often introduced by a character called Jokhim Chacha (Joking Uncle) whose age the audience are asked to guess.

Figure 10 SWRC, solar cooker workshop

As they increase their estimate of years his neck grows to about half a metre until they reach his age, which is 365. But Uncle is not a joke, more a joker who can say what actors might not voice, and who, because he is so old, is utterly respected. If he advises a form of disease prevention, for instance, or improvements in maternity care, or that women should be respected, he will be listened to. The puppet workshop offers training in making puppets and giving performances to members of other cultural groups in Rajasthan as well as local villagers. Currently a team of four puppet-makers and performers works with two villagers as trainees, and has trained about 2,750 barefoot communicators while performing in more than 4,000 village locations.

Alongside lampooning government officials, a serious issue is access to information on how money granted for local development has been spent, or if it has been spent at all, while a predominant lack of accountability tends to produce corruption. Street theatre and puppetry encourage dialogue and political as well as social awareness, in the context of other, direct means of village self-organization. For example, faced with a lack of transparency in development, in the early 1990s, the Mazdoor Kisan Shakti Sangathan (MKSS, or the Labour Farmer Strength Organisation), formed by villagers, walked from village to village to ask about development funds and accounts. Not surprisingly at the time they were told they had no right to such data while all receipts, vouchers, musters and so forth were not available. MKSS launched a campaign with public hearings on corruption and the misdirection of funds, and mounted a 53-day strike in Jaipur which produced a state order to make available the information requested – with limited results at the local level. MKSS then returned to the villages to publicly challenge local officials and local assemblies to be honest about where money had gone. In some cases corrupt officials returned funds they had misused: 'No law or arrest or department enquiry made these officials return the money. It was the local villagers who finally humbled them' (www.barefootcollege.org). SWRC sees such action as part of village empowerment, calling on non-governmental agencies to adopt similar policies. It has itself, since 1997, shared all records of its income and expenditure at public meetings.

The concern for human rights extends to women in a male dominated society, who are empowered through training in solar engineering, water harvesting, literacy, village-level education, internet use and other programmes. I saw village women using laptops in the internet hut, a geodesic dome beside the administrative building (Figure 11). One tactic for advancing women's rights is the formation of women's groups to shift the attitudes of village men. SWRC has also opened personal bank accounts for women engaged in craft work in the textile workshops at the Fever Hospital; their products are sold in the Barefoot College craft shop and, through it, to external agencies such as Oxfam. The key point is that the money raised goes directly to the women, who exercise control over its use. I was told that the men would use the money to buy illicit liquor (in a dry state) and have sometimes been unable to work for two or three days after a binge-drinking bout. Many women have set aside money for their children's further education, or to buy equipment such as a sewing machine as a further income generator.

Figure 11 SWRC, geodesic domes and administrative building

Health care and night schools

Finally, I want to look at health care and the provision of night schools, both central to SWRC's work. Set up in 1973 with a curative dispensary in a geodesic dome, the health care programme now extends to working with villagers on a range of health issues but with a particular emphasis on pre- and post-natal and maternity care, and has led to the training of barefoot health workers able to dispense bio-chemic medicines (for which formal qualifications are not required, and which have no side effects) in the villages of the surrounding area. Traditional midwives come to the college to upgrade their skills, and young people (male and female) are selected by village organizations to receive residential training in health care and pharmacy work. Training covers HIV awareness, nutrition, anaemia, eye care, mental health and hygiene as well as maternity and pre- and post-natal infant care. Much of the work on health care is preventive medicine, such as improving birthing practices and attitudes to women and infants in a society where most women work from dawn to dusk in fields right up to giving birth. The programme of pre- and post-natal care has dramatically reduced infant mortality rates, from 75.6 infant deaths per 1,000 live births in 1999–2000, to 56.7 in 2000–1 and to 17 in 2001–2, against a regional average of 94 (in Ajmer district) and a national average of 87 in 1997. The actual numbers are 83 infant deaths in 1,096 live births in 1999–2000, 51 in 899 in 2000–1 and 25 in 1,473 in 2001–2. Maternal deaths in childbirth were 1 in 1999–2000, 2 in 2000–1 and 1 in 2001–2, again against much higher national figures. The reduction in infant mortality means that, in turn, the pressure to produce a large family as a source of agricultural labour is reduced, complemented by advice to women on birth control.

The three key dimensions to the health programme are health education, preventive medicine and treatment, through training for women and boys who have an interest in health work and have completed their schooling. Survey work is carried out to monitor health needs and outcomes, and information is exchanged with government health services (which have less contact with villagers). The programme seeks to demystify health care at village level, linking modern medicine to village knowledge through personal contact. An increasing problem is the incidence of tuberculosis contracted by villagers working in the nearby marble quarries, usually after four or five years of this unsafe and badly paid employment in an industry largely supplying foreign markets at prices below those of most other sources. When workers become too ill to work, health care can alleviate the symptoms, perhaps, while only the empowerment of villagers will change the conditions in which they arise.

During my visit to the Barefoot College in 2004, I was taken to two night schools. At the time of my visit there were 147 such night schools operating (since increased to more than 200) in which there were a total of 1,479 boys and 2,736 girls – a major contribution to gender equality. The schools are free to any village child, and those that participate are aged from about three to about 15. All the teachers are barefoot teachers – villagers who have trained at the Barefoot College, which produces and prints its own training books and materials. Today there are satellite night schools in Bihar, Orissa, Madhya Pradesh, Uttranchal and Assam, as well as throughout Rajasthan, as teachers trained at SWRC have gone back to their states. As said above, all night schools use solar lanterns. Each night school elects a representative to a Children's Parliament. The Parliament is provided with a secretary by SWRC, and has the power to change the curriculum (or the teacher if deemed necessary).

The reason for night schools is simple: the government provides daytime schools, but in rural areas most children, especially girls, go to work in the fields with their mothers for the day, and cannot go to the government-provided school. Night schools have proved effective in increasing access to education for girls – who have traditionally fetched water in the evenings when the schools are held, often walking long distances to do so – because water harvesting centres on the village schools and community buildings in which the night schools are held, so that girls are able to access the water supply when they participate in the night school. Sessions usually last about two hours per evening, and cover literacy – beginning with each student learning to write her or his name in hindi – mathematics, basic animal husbandry and other skills used in village life, and basic science. It was explained to me that the students also learn about society. I asked what this entailed. The reply was that, by way of illustration, some of the older children (which means young adults of 14) work in the quarries. They said the foreman paid them a daily rate of 35 rupees but wrote down that he had given them 45 rupees, presumably pocketing the difference, unaware that the young workers were literate and numerate and would feel empowered to act on his corruption. The matter of quarry wages was discussed at a session of the Children's Parliament, from which a case was made to the authorities as a result of which, after some negotiation, the rate paid was increased to 60 rupees, the legal minimum wage. The difference is, in Western terms, less than the cost of a cup of tea, though in Rajasthan it will buy a bag of fruit or vegetables at a local market and could be a critical amount to the family supported by the quarry labourer. But to me the main significance was that the students had learned solidarity as a basis of self-organization in support of equity and human rights – that is, they had learned that by acting together they could change the world (or at least prevent an injustice). This was relayed to me by the students themselves (through translation) in the light of three or four solar lanterns while we all sat on the ground (Figure 12). I was relieved they did not ask me what equivalent education for empowerment is available to children in Britain. I do not think classes in civics at secondary school are a match for this, and perhaps a first degree in sociology is not either.

The night schools fuse education and direct democracy and are likely to be of long-term impact in empowering the rising generation of villagers, perhaps enabling them to remain in village society and assist its appropriate development rather than migrating to cities where conditions can be considerably worse. Just as all the teachers are villagers who have completed residential courses at the Barefoot College, so each village has a Village Education Committee of 15 to 30 people, at least five to 10 of whom are women. The committees manage the night schools' funds through rural banks, purchase teaching materials, pay honorariums to teachers and deal with repairs to premises. They meet monthly, and encourage parents to send children to the schools. But, as the Barefoot College website states:

Figure 12 SWRC, children at a night school

Children... monitor their own schools, by electing their own representatives.

The Children's Parliament controls and supervises the night schools. It is based on the belief that giving power to the people who have a vested interest in the school is the best way of ensuring its success – as well as making the children aware of political structures and processes.

This form of education-related activism provides a heightened awareness of the system, its workings, and avenues for redress of local grievances.

(www.barefootcollege.org)

I left Tilonia feeling my academic qualifications had little to offer in village India, and have written this book partly in gratitude for the six days I spent at SWRC. I have said little of the buildings beyond that they were built by barefoot architects, the social architectures seeming more important (while barefoot building is a social architecture). In fact, SWRC adopted the geodesic dome from the West, as a simple structure that could be made, like the framework for a solar cooker, from scrap metal, then covered in a range of materials from plaster to thatch. The domes are light enough to be put on a cart and taken by tractor to a village as a medium-term structure.

Abjection and Utopia?

This leads me to wonder if there are elements of SWRC's work that could be adopted in the West. In terms of the social architectures of SWRC, the question is not only what tactics can be learned from the programmes described above but also who, in an affluent society, are the poorest of the poor with whom to work. Before suggesting one of several possibilities in response to the question, I want briefly to relate the work of SWRC to other currents in post-colonial discourse, identifying issues of abjection, empowerment, voluntary simplicity and utopian desire. It is easy to say these are, variously, opposites or that they complement each other, though a form of empowerment is implicit in a life of voluntary simplicity, while Utopia is likely to be a society in which none are abjected. The difficulty, in relation to the flaw in Fathy's efforts to create a mud-brick Utopia, is that abjection – the villagers' lack of a means to improve their own conditions – does not produce what in the West would be seen as a utopian desire. More often, as in the North American desire for wilderness, it arises in an absence, as for wilderness in an urban-industrial society, so that laments for the loss of a mythical pastoral life are framed not in a future to be built but in a loss perceived as a loss of wholeness or social cohesion. To map this onto a non-affluent, post-colonial society is inappropriate, redirecting attention to the failure perceived by Fathy, both in his project at New Gourna and in the peasant class, to be able to go beyond abjection. That is, the distancing of the utopian desire or vision itself impedes efforts, not to lift the peasant class out of their state as it is, but to recognize that in that state exist certain knowledges, skills and forms of organization that may be a basis for empowerment. Fathy realized this in terms of buildings but not in the equivalent terms of social architectures or a grassroots culture.

Frantz Fanon writes in *The Wretched of the Earth* (1967) that the national consciousness of a colonized people is likely to be, as a result of inscription on the colonized society of an alien culture, 'an empty shell, a crude and fragile travesty of what it might have been' (Fanon, 1967: 119). The colonial view erases differences within the colonized culture so that 'the nation is passed over for the race, and the tribe is preferred to the state. These are the cracks in the edifice which show the process of retrogression . . . so harmful . . . to national effort and unity' (ibid.). Fanon continues that the middle class of the colonized state, a client middle class, tends to mimic the values of the departing colonial power. Then, 'The university and merchant classes which make up the most enlightened section of the new state are . . . characterized by the smallness of their number . . . being concentrated in the capital, and the type of activities in which they are engaged: business, agriculture, and the liberal professions' (Fanon, 1967: 120). These are the classes who assumed influence in post-independence India, and established the social work tradition from which SWRC departed. Fanon writes in the context of French Africa, where conditions differ from those of India, yet a common question occurs as to the construction of a national cultural identity. Leopold Senghor and Aimé Césaire, among others, evolved the concept of negritude as a positive image of the black-skinned person called the Negro, to refuse the abjection donated by the colonial power. In India, an Indian administrative class emerged in the nineteenth century within colonial rule, some of its members becoming social reformers and educationalists. Jyoti Hosagrahar writes:

> Delhi acquired a new group of indigenous elite. Indians schooled in the British system of education joined the colonial bureaucracy. Members of the new cadre were English-educated, 'reformed', and in the service of the Empire as magistrates, treasurers, accountants, and civil servants.
>
> (Hosagrahar, 2005: 35)

The cultures of the British regime and Indian elite society were linked, and in the desire for a better post-independence world the regulation and hierarchy that characterized the colonial regime were retained. Gandhi's voluntary simplicity was revolutionary in that the means embodied the ends desired, while the attitude of India's middle and professional classes to the poor remained a relationship of disempowerment even when seeking improvement in their conditions. This goes part way to explaining why some housing schemes (Chapter Eight) have failed to create better conditions for the economically weak sector: well-intentioned, enlightened and progressive ideas framed in a Western utopianism, and not extending to appropriate social architectures by which members of the peasant class can empower themselves, do not deliver that empowerment. Fanon writes that 'only when men and women are included on a vast scale in enlightened and fruitful work' is a national cultural identity able to evolve appropriate forms, and continues: 'the flag and the palace where sits the government cease to be the symbols of the nation. The nation deserts these brightly lit, empty shells and takes shelter in the country, where it is given life and dynamic power' (Fanon, 1967: 165).

Frampton uses the term Utopia in his gloss on SWRC, which I reiterate more fully here: 'In the last analysis, the Barefoot College displays all the traits of a Utopian community abstracted from another moment in history' (Frampton, 2001: 12). Likening the campus to a Fourierist phalanstery, he cites the open-air stage flanked by 'the cultural and collective core of the college, . . . the main dining room, the puppet theatre and administrative offices' (ibid.). Frampton's point seems to be that the cultural life of SWRC takes the place of the parade ground of the phalanstery. Does Frampton want simply to know how to read somewhere as strange to him as the Barefoot College, falling back on a convenient Western paradigm? I do not want to dwell unduly on his remarks, but they draw out a habitual Western alignment of utopian settlements to fantasy or severity – the dream-book or the monastery.

SWRC, it appears, achieves a rupture of such utopianism by breaking the vicious circle of a picture of freedom framed in a structure of values derived from colonialism, and recognizing that the basis for a new society already exists in village societies. There is professional expertise, and in fields such as health care, or in facilitating the project's establishment and funding, it is of course essential; but the point is to hand over power and give up the mystification of expertise and consequent dismissal of the tacit, experiential knowledges of dwellers that expertise entails.

Empowerment and voluntary simplicity

For SWRC, identification with village societies was vital to bridging the divides of class, gender and urban/rural status. In demystifying expertise, and in listening to those whom the intervention is designed to help so that they can frame it rather than receive it, SWRC has empowered villagers. Listening is a means to contribute to the conditions in which others can empower themselves, just as talking at people as if they were a wall has the opposite effect. In this respect I cite the work of Paolo Freire, Brazilian educationalist and author of *Pedagogy of the Oppressed* (1972), in which he sets out what he calls the banking concept of education in which the teacher and the student are divided by the teacher's expertise. He writes that analysis of this relation 'reveals its fundamentally *narrative* character' and that the contents of education, whether values or empirical findings, 'tend in the process of being narrated to become lifeless and petrified' (Freire, 1972: 45). He continues:

> Education thus becomes an act of depositing, in which the students are the depositories and the teacher is the depositor. Instead of communicating, the teacher issues communiqués and 'makes deposits' which the students patiently receive, memorize, and repeat. This is the 'banking' concept of education, in which the scope of action allowed . . . extends only so far as receiving, filing, and storing the deposits. . . . Knowledge [in contrast] emerges only through invention and re-invention, through the restless, impatient, continuing, hopeful inquiry . . . with the world, and with each other.
>
> (Freire, 1972: 45–6)

Knowledge, like power, or as a means to power, is either an element in a top-down and disempowering development or is created by the learner. This sounds quite abstract and it should be remembered that in colonial societies a mark of power is the ability to write, to be a member of the lettered society. Freire responded to this inherited state of affairs by asking students to tell their own stories rather than following a curriculum. He writes:

> The truth is, however, that the oppressed are not marginals, are not men living 'outside' society. They have always been inside – inside the structure which made them 'beings for others'. The solution is not to 'integrate' them into the structure of oppression, but to transform that structure so that they can become 'beings for themselves'. Such transformation, of course, would undermine the oppressors' purposes; hence their utilization of the banking concept of education to avoid the threat of student conscientization.
>
> (Freire, 1972: 48)

From this, 'the humanist, revolutionary educator... [can] engage in critical thinking and the quest for mutual humanization' (Freire, 1972: 49).

Freire's model of education emanated in a non-affluent society, in a simplicity or voluntary humility on the part of the teacher, that I would equate with the voluntary simplicity of the work of SWRC. It is not a hair-shirt, self-demeaning impoverishment (likely to be involuntary if it proceeds from an unconscious desire). It is the condition of living in what, through careful analysis of conditions, is an appropriate and sustainable way. It is, as Freire's approach draws out, through such simplicity that power-over is dissolved because it has fewer objects, while those who lack power-over find they have power-to just as those who had power-over are liberated, too, from its alienation when they relinquish it.

Today, in the affluent West, the question of simplicity is inflected by global warming to become a search for ways to reduce the footprint of living on an overloaded – which is not necessarily to say an overcrowded – planet. It may be inflected by a residual dualism from the Cold War as well, of an affluent West and an East that is neither affluent nor non-affluent (neither First nor Third World in the language of the day) but frugal. The Utopia envisaged by socialist and anarchist revolutionaries in the nineteenth century, and exemplified in the principle that each person works according to ability and consumes according to need, is in its way a form of voluntary simplicity. Victor Papanek's call (1984; 1995) for a revision of wants according to more sustainable ideas of need in human societies carries a similarly voluntary simplicity. But if the state socialism of the post-1917 period in Russia, and post-1945 period in Eastern Europe, had any vestige of this, it was reinterpreted, and I would argue misrepresented, by the West during the Cold War as a negative, imposed frugality. From the 1970s, consumerism began to encroach on this world, and May Joseph writes of Hanoi, a post-colonial city that was also a city of the Second (state socialist) world:

> For socialist cultures of the last century, the frugal city bears a special place in the Second World imaginary ... filled with the ghosts of its history, and

the spectres of 'ours' . . . The frugal city was a utopic city, what Foucault calls 'a site with no place'. It was the space of the ideological phantasmatic, a city quietly resilient in its conviction to survive at all odds.
(Joseph, 2002: 246)

Joseph reads a frugality into the aesthetics of Western environmentalism, 'invoking images of a depleted earth in the interests of planetary sustainability' (Joseph, 2002: 247), adding that the same quality affects nineteenth-century North American utopian communities such as Brook Farm (see Francis, 1997). Perhaps frugality always has that negative dimension. Perhaps it is difficult in the West to imagine simplicity other than through images of medieval monasticism or some form of punitive self-denial. Within the affluence of a consumer society, however, is an abjection of those who are not party to it, who lack the mobility of the new elite as well as their power of consumption, among them asylum seekers and political refugees.

Coventry Peace House

To accept an impossibility of voluntary simplicity in the affluent society would be to deny the possibility of empowerment to live in ways that are not dependent on the fetishes of commodity and consumption. The evidence, in any case, is that there are communities practising voluntary simplicity within the affluent society, such as Tinker's Bubble in Somerset (Chapter Seven) whose members use no fossil fuels. I end this chapter with a case of a commune in the English West Midlands working with asylum seekers, because asylum seekers are among the poorest of the poor in this society, marginalized in multiple ways and for the most part denied a basis for solidarity by being isolated from the social structures of their background. Asylum seekers receive little in the way of state benefits while being barred from employment while their cases are deliberated over; in some cases they are told, when their last appeal fails and the state has no reciprocal arrangements with their home country or means to transport them to it, to find their own way home. At this point they are stateless people with no means of support, sourcing the statistics of people who have vanished into unknown circumstances.

The Coventry Peace House is a group of six terraced houses in Coventry. Its name reflects the fact that it was founded in January 1999 after a 13-month camp outside a local arms manufacturer's premises. It operates as a mutual housing cooperative with six members. The houses have been renovated to provide individual living accommodation and spaces for meetings, common dining and community work. The kitchen is the central space of the community where meetings tend to be held and visitors received, but there is also a common sitting room. There is a large garden, too, and plans for grey water recycling, solar power, peace education and a neighbour mediation service. The coop is based on the principles of non-violence and mutual aid, and potential members have a three-month induction period. Work is organized on a voluntary, shared basis, and each member has his or her own interests and in some cases employment. All decisions affecting the cooperative are made by consensus at weekly meetings with an agenda and minutes, which can last as long as four hours. As in other intentional communities (Cases Seven to Nine) the time spent in meetings is considerable and items that take time to decide can seem trivial to an outsider; but the process enables anything to be voiced, whether seemingly on the subject or not, and including feelings about the actions of other members. As one member said, everyone has different reasons to be there and different agendas for the group; weekly meetings are where these differences are worked through with the assurance of privacy but also honesty and directness. Another member said the desire is for each person to be fulfilled but not in a conventional way, interpreted as personal satisfaction or power over others. Among matters recently raised at meetings were whether to compost food waste, which attracted rats to the compost heap; this took two hours to discuss. Among other matters already decided are that the group is vegetarian but not vegan, non-smoking, and has no televisions or cars.

A difficulty in consensus decision-making can be that it breeds conservatism and a tendency for things either not to happen or to be postponed. At the same time, there is a sense of mutual respect for the different skills possessed by each of the current members (including house renovation, art, music, cycle repair, the tactics of non-violent resistance and sociology). The lifestyle is simple but not frugal, with a commitment to the shared ideas of a self-organizing society serving the needs of others as well as of its members.

I visited the Peace House in October 2004, when it was working with asylum seekers more than on peace issues. This involves collaborating with local, national and religious groups to assist asylum seekers in writing letters, making new contacts, meeting people of their own language group, understanding the bureaucratic answers they receive from government offices, dealing with appeal procedures, and at times providing shelter and close support, or a social space. Currently, in reaction to tabloid press campaigns against asylum seekers or foreigners in general, government restrictions on supporting asylum seekers whose appeals fail make such work more difficult, to the point that some church groups feel legally bound to withdraw at a certain stage. But prior to that point Home Office funding is available, and supports such work at the Peace House.

Among the ideas discussed in conversations during my visit – only for a day – was immanent logic. Immanence, a pervasiveness in the present, was felt to be an alternative to instrumentalism (the model in which means are seen as delivering ends rather than as processes in their own right). Immanent logic was seen as avoiding habitual reproduction of conventions regardless of the stated policy or objective, and I would see it as taking the means not as a signpost to the envisaged ends but as the ends themselves, inasmuch as the means always produce the values they enact. Another theme of conversation was the evolution of a lower-impact lifestyle through alternative energy and a redefinition of needs. My final point is modest: that this is revolution by other means, paradoxically in the immanence of a commune and the imminence of working to deal with the dominant society's casualties. I resume discussion of this in the book's Conclusion, after the nine short case studies that follow this chapter.

Part Four
Short case studies

Case 1
Economy, Pennsylvania, USA

Economy Village is in Beaver County, near Pittsburgh in the hills of Pennsylvania. The site is preserved as an open-air museum, with access to the interiors of several buildings including the Feast Hall, a cabinet shop, a granary, a warehouse, a merchants' building and the houses of George and Frederick Rapp among others. The larger buildings are in red brick; others are whitewashed and all are well crafted and have lasted well despite the abandonment of the site in the 1900s. The interiors of the houses are plain but light, still furnished as they would have been in the period when Economy was a thriving religious utopian settlement. But the buildings are in a time warp, an outdoor museum minus its living inmates. Perhaps Economy was itself in a time warp for the later part of its existence, as numbers dwindled and the remaining residents were forced to hire labour from surrounding areas to carry out the work of maintaining its industries and its agricultural production. A

Figure 13 Economy, the community kitchen

guidebook states 'Tours of Old Economy Village begin at the Visitor Centre, with the purchase of admission tickets' (Reibel, 2002: 29). There are the usual guided tours, or one is free to wander alone. The museum site is in public ownership, though a few houses in adjacent streets beyond the original village seem to be of the same period and type of construction. A more recent settlement, now called Ambridge, has grown up around Economy, and Pittsburgh, with its redundant steel mills and slowly rolling freight trains on countless waterside tracks, is about an hour's drive away. Is this what happens to all utopian settlements?

In 1785, George Rapp, a vine-dresser and linen weaver from Württemberg in south Germany, founded a religious congregation as a departure from the Lutheran church. The latter he perceived as corrupt, though Lutheranism was itself a departure from the Roman church on similar grounds in the sixteenth century. In 1804 Rapp bought 4,000 acres of land in northern Pennsylvania, and began to build log houses, the beginning of the agricultural settlement of Harmony. The name Harmony was adopted by Rapp before Fourier used it for his ideal community, and it is unlikely that a copy of Fourier's treatise published in the *Bulletin de Lyon* in 1803 would have been available in Pennsylvania (see Chapter Three). Brick buildings followed, and a fulling mill for cloth. In 1805 the Harmony Society was founded, its members – mainly from the peasant and artisan class – donating all their property to the society's ownership for the collective good. In 1807, the society adopted the practice of celibacy for religious reasons (though a few children were born into it, and some families left at this point). Mostly, it seems, husbands and wives lived as brothers and sisters in Harmony, sleeping separately but working together. The long-term result, of course, was that the number of members declined. One commentator suggests that women became less feminine after the decision on celibacy, being assigned men's jobs and adopting a severe appearance (Larner, 1962: 215). This may be negative propaganda, however. A traveller in 1811 records that Harmony was laid out in regular proportions, with a central square, three east–west streets crossed by three north–south streets, subdivided into quarter-acre lots, like a city in miniature. Each family had a house, two cows, pigs and poultry (Melish, 2003: 15). This ordering is typical of the settlements planned by nineteenth-century utopians – Robert Owen's vision of a model community (as drawn by Stedman Whitewell; Lockwood, 1905: facing p. 70) has this regularity in a square plan of buildings with regularly proportioned paths across its interior spaces.

Manufacturing and agriculture prospered at Harmony, cloth being the main trade. In 1814 the Harmonists sold their settlement to a Mennonite group and moved to Indiana, calling their second settlement Harmony as well. Recruiting in post-Napoleonic Europe, capitalizing on a time of ferment and insecurity, the society grew to around 900 members, all of whom enjoyed equality after records of original donations of property were intentionally destroyed in 1816. Cloth was produced in steam-powered mills, and sold as far away as New Orleans and Philadelphia. But Indiana was a long way from sources of raw materials, while the Harmonists got on badly with their neighbours and suffered fevers in the damp heat. In 1824 Rapp's adopted son Frederick found a 600-acre site for sale near Pittsburgh, which was purchased. By 1825 a majority of Harmonists moved to Economy, as the new site was named. Harmony was bought by Owen as a proto-site for a new community based on principles of education as a means to mass social improvement (Chapter Three). Owen renamed the site New Harmony. His son, Robert Dale Owen, notes: 'Harmony was a marvellous experiment from a pecuniary point of view, for at the time of their emigration from Germany, their property did not exceed twenty-five dollars a head, while in twenty-one years (i.e. in 1825), a fair estimate gave them two thousand dollars for each man, woman, and child, probably ten times the average wealth' (Lockwood, 1905: 28). George Lockwood adds, though, that the price paid by Owen was half the site's worth. Lockwood also doubts accounts (based on Charles Nordhoff, 1966) that Rapp and his community left due to fever and bad neighbour relations, suggesting market reasons – to be nearer the urban centre of Pittsburgh – were more important. The life of Economy continued as that of Harmony had done, the village being built and the new estate completely paid for within eight years. A steamboat was commissioned to transport the community to the new site.

Substantial industries were established, and regular feast days held. Given the rule of celibacy, the feasts held in Economy are unlikely to have resembled those of Fourier's eroticized fantasy world:

> The Harmonists held feasts about four times a year . . . The Love Feast, or *Liebsmahl*, was a common practice in pietistic societies. . . . The feast included a meal and a religious service and lasted all day. . . . Little is known about how the Love Feast was conducted, as outsiders were never invited. Prior to the feast, the members would confess all individual differences before father Rapp. If their feast was conducted like those of other pietistic societies, it would have included the ritual of participants giving each other the 'kiss of peace' . . . Once all were in 'the harmony' they would go into the feast.
> (Reibel, 2002: 30)

Charles Nordhoff saw 'Neatness and a Sunday quiet' as the prevailing characteristics of Economy village, which he visited in 1874. He reports the arrival in 1831 of a German adventurer, Bernhard Müller (calling himself Count Maximillian de Leon) and his group of followers, who then began to announce strange doctrines. In direct contradiction of the Rappites' celibacy, Müller proposed 'marriage, a livelier life, and other temptations to worldliness' (Nordhoff, 1966: 79). He persuaded some Harmonists to side with him, and the dispute became bitter, resolved only by a vote in which around 500 supported Rapp and 200 Müller. Nordhoff remarks that on hearing the result, Rapp, 'with his usual ready wit, quoted from the book of Revelation, "And the tail of the serpent drew the third part of the stars of heaven, and did cast them to earth"' (Nordhoff, 1966: 80). The outcome was an agreement to buy out Müller's supporters on condition they left Economy, which they duly did, going to Phillipsburg and there setting up a new community allowing marriage. Müller absconded in 1833, and died of cholera. The new community failed.

182 *Short case studies*

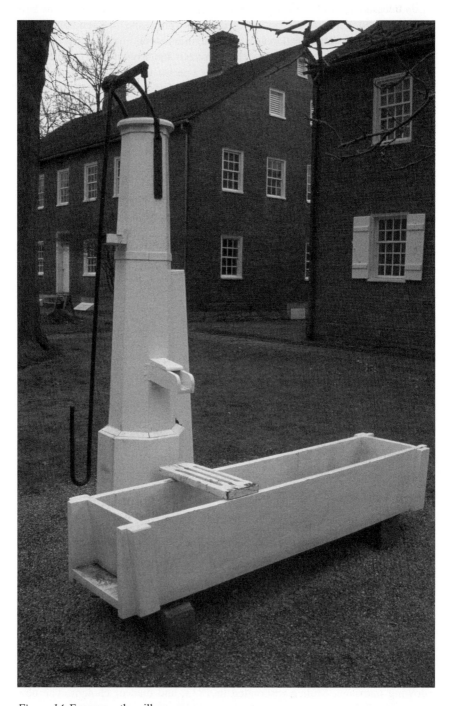

Figure 14 Economy, the village pump

By the time of Nordhoff's visit in 1874, celibacy had taken its toll at Economy. A dwindling and ageing population was dependent on hired workers to maintain the industries and agricultural production, and no longer able to manage Economy as the highly profitable business it had once been. Nordhoff writes, 'Once it was a busy place, for it had cotton, silk, and woollen factories, a brewery, and other industries; but the most important of these have now ceased' (Nordhoff, 1966: 64). In 1874 there were 110 inhabitants. Although some children were adopted by the society, Nordhoff reports that it did not seek new members, withdrawing into itself. In 1892 there were only six Economists left. The area surrounding the village's original centre was sold in 1902 for $4 million to the Berlin Iron Works (later the American Bridge Company, after which Economy was renamed Ambridge). The last Senior Trustee (effective head of the Society), John S. Duss, had, since the late 1890s, developed an enthusiasm for brass band music, in 1903 using funds from the sale to support his ambition to take the band he had founded on tour. He hired the Metropolitan Opera Orchestra and had an imitation of a Venetian canal made in New York's Madison Gardens – with gondolas – for his performance as conductor. A guidebook states: 'Duss equated himself with John Philip Sousa, but the critics were less kind. The concerts ended in 1903, though the Economy Band continued to exist through 1905' (Reibel, 2002: 26).

Case 2
Arcosanti, Arizona, USA

Arcosanti, near Cordes Junction in the canyon country of Arizona, is described by its architect Paolo Soleri as a potential urban laboratory (Soleri, 1993). In an appreciation of Soleri's work, John Cobb Jr. writes that civilization is heading for disaster, and that 'much of the problem is caused by the way our cities are built' (Cobb, 1993: 9). The foil to this disaster script is arcology, Soleri's approach to urban design and building which Cobb characterizes as a sustainable 'consummation of the human adventure' (ibid.). Arcology is defined on the back cover of *Arcosanti: An Urban Laboratory?* as 'a methodology that recognizes the necessity for radical reorganization of the sprawling urban landscape into dense, integrated, three-dimensional towns and cities' (Soleri, 1993). Jeffrey Cook summarizes Soleri's intention as honouring 'man [*sic*] . . . the single surviving primate, distinguished by self-conscious thought processes coupled with con-

Figure 15 Arcosanti, the city on a hill

templative and engaging intellectual prowess'; and fusing *homo sapiens* with *homo faber*, the maker (Cook, 2001: 11). He roots Soleri's view of the city in Mediterranean urban life, with its 'gregarious play of humans, in which audience and actors are the same' (ibid.). This image of everyday life is backed up by reference to Greece and Rome, and Soleri is credited with a desire to propagate universities and art galleries as 'markers of civilization' (Cook, 2001: 13). Cook places Soleri in a tradition beginning with Frank Lloyd Wright (for whom Soleri briefly worked, until he was fired) and Le Corbusier, and characterized by Le Corbusier's plan for Algiers as 'liberation from the land' (ibid.; see also Chapter Four).

Yet, as Cook points out, Soleri refused to be associated with the urban gigantism of Reyner Banham's book *Megastructures: Urban Futures of the Recent Past* (1976), claiming instead the virtue of miniaturization. Perhaps it is unfair to burden Soleri, far less the present residents of Arcosanti, with this background in Modernist architecture, though the reference to Le Corbusier may be illuminating inasmuch as Arcosanti is an attempt to create a society engineered by design, not entirely unlike the cell of which Le Corbusier writes (see Chapter Four), made new on top of a hill. It clearly cannot be read in the Orientalist context of Le Corbusier's Obus Plan for Algiers, but perhaps the North American West gives rise to an Occidentalism that is as much a mythicization of the material histories involved, in this case of white settlers creating a new world in what they necessarily took as an unpeopled land.

Arcosanti was intended as a pilot city for 5,000 people. Soleri describes the derivation of the project after living in Santa Fe – known for adobe building (McHenry, 1996: 108–12) – and previous urban design efforts:

> So I came up with this idea about a city on a mesa [hill top]. I really was taken by the feeling you have when you're out on a mesa, by the beauty of the landscape . . . But Mesa City . . . was too big. It was too flat. It was designed to hold a million people in a space about the size of Manhattan. . . . as I was working out my ideas . . . I found out that it would need transport systems . . . which are fatally overwhelming. So I saw that something was screwed up there. At the same time, I had this insight, this eureka, about exploring the implications, in the design of a city, of complexity and miniaturization.
>
> (Soleri, 2001: 34–5)

Arcosanti is an intentional community, currently home to around 70 people which Soleri visits weekly from his ranch; it is also a continuing experiment, in which most of the construction work is done by students during summer schools. The students occupy one-room wood and perspex cubes in the valley below the city's hilltop site, next to the fields where Arcosanti's vegetables are grown. The settlement covers around 25 acres, and has a large central arena used for music and other public events, surrounded by terraces. There is a small library used as a study centre, apartments for residents, a terrace of rooms for visitors overlooking the canyon, an open-air pool, vegetable gardens, a grey water recycling pond, and facilities including kilns and casting equipment to produce the ceramic and bronze bells that provide Arcosanti with its main income stream, which I was told is $1 million a year. The bells are expensive – in bronze up to several hundred dollars – and are the most common purchase made by visitors (though not by me).

Soleri began to make slip-cast bells in Santa Fe, though the climate was too cold. Hence Soleri's move to Scottsdale, Arizona, from where he still oversees the work at Arcosanti. From his experiments he found concrete could be slip-cast in a way similar to that used for clay, and used this method to make the arches of Arcosanti. The settlement initially consisted of a few basic shelters in the canyon (like the student huts) before work began on the mesa. Today there are rows of apartments around the central auditorium and in several blocks elsewhere. Soleri states that the form of the settlement is informed by frugality (Soleri, 1993: 39; 2001: 41–2), but that it does not aspire to self-sufficiency (in contrast to many intentional communities founded in North America in the 1960s). He writes: 'via complexity-miniaturization and the discipline of frugality, a more restrained and judicious use of resources from near and far (Sun, 93 million miles away) will favor a condition of self-sustenance' (Soleri, 1993: 39). Today the absence of solar and wind power generation seems at odds with this, though one resident suggested that while Soleri was not interested in it, younger members of the community might take up such ideas at some point. Arcosanti produces most of its food from adjacent fields and organic gardens, however, served in the public restaurant daily, and has a grey water recycling scheme for irrigation.

Although the present size of Arcosanti is small, its design and material construction are essentially urban: 'One thing that has always irritated me is that there has been no hint in the environmental movement . . . about the connection between our presence in terms of habitat, and the environment. I have always said that the salvation of the environment is in the city' (Soleri, 2001: 44). Arcosanti, though, looks from a distance (Figure 15) like an Italian hill town with its cypress trees (an imported species, needing to be constantly watered by sprinklers in the hot sun of Arizona). There are no roads, only footpaths and several wide public spaces; those who gather in these, however, are mainly visitors while residents keep a low profile, inured to but no longer welcoming the burden of living in a visitor attraction and dutifully giving guided tours of their settlement according to a more or less agreed and at times more interestingly improvised script.

Bells are produced in ceramic and bronze, with kilns and a foundry on site, to fund the project. There is always a construction site somewhere because the pace is slow and the work likely to remain unfinished. Yet, despite its imposing visual presence as a city on a hill, seen from over the canyon and the usually dry river bed, Arcosanti is in its present state on the scale of a small village. This, as it happens, is the typical scale of an intentional community or ecovillage (Chapter Seven) and might as such facilitate consensus decision-making as to how the project should develop in future. Instead, though, at present there are weekly question and answer sessions when Soleri makes his visit.

I found Arcosanti an uncanny Utopia, a city planned in completeness in which successive dwellers might add detail but only within the confines of the original dream they are required to manufacture. I wondered at the ethics of using students

Figure 16 Arcosanti, the Modernism of cypress trees and concrete

for construction work (though it is voluntary, they even pay for the experience); and found few residents wanting to talk about the mismatch of the plan for a convivial city for 5,000, meeting the needs of minds and bodies, and the diminished size of the community today. I enjoyed being there in the simple, en suite guest rooms on the edge of the hill, looking out over a landscape I had seen previously only in films. I swam several times a day in the open-air pool that no one else seemed to use. I left quietly, walking across the desert to the gas station where a taxi would take me to Phoenix, a city with nothing of the idea of a city about it. I felt very European in this very North American urban landscape.

Case 3
Auroville, Tamil Nadu, India

Auroville is an international city in Tamil Nadu, southern India. The plan for the city, initially conceived for 50,000 people, is overlaid on the map of several Tamil villages occupying a low-lying plateau. At its centre are gardens, an amphitheatre and a gold-domed temple called Matrimandir. The plan demarks an inner circle (city area) around this containing public buildings, and an outer ring (green belt). From the centre, four zones spiral to the edge of the area marked for building the city, for culture, industry, housing and international pavilions. Spread irregularly across the map, in both the city and the green belt, are clusters of houses like hamlets, constituting a more informal or non-planned city built as Auroville's residents acquired land and built or commissioned houses. The names of these residential clusters include Existence, Sincerity, Surrender, Invocation, Adventure, Creativity, Acceptance, Grace, Gratitude, Bliss, Fraternity, Gaia, Dana, Sangha and Transition. The role of Auroville's urban planning department, called Auroville Futures, is to reconcile this informal development (though not a development of informal architecture, since most of the houses reflect modern European or Indian styles of architecture in brick or concrete) with the need for a coherent infrastructure and provision of services.

At present Auroville houses around 1,700 people, from those who arrived in the 1960s and are now in their sixties or seventies (beginning to receive elderly care within Auroville) to others recently arrived, mainly in their thirties. A larger number of Tamil villagers work in Auroville. The majority of residents are European or North American expatriates, with some Indians. There is no binding rule, as in a monastic community, but Auroville has an induction process. Commitment is required as well as acceptance by an Entry Group of Aurovillians. A guide booklet emphasizes the difficulties of living there: 'It's not just the heat . . . nor the sensory and emotional overload, nor the galloping entropy that causes almost everything to fall apart very quickly . . . everybody who comes to live here seems to be confronted . . . with difficulties . . . rarely experienced before' (Auroville, 2002: 6). As the booklet explains, progress is impossible until the problems that compose a person's psychic and somatic baggage are faced, and then gone beyond. Perhaps the same applies to the city as to individuals living there: in some cases they are engaged in collective endeavours, in others working in small groups on local projects for land reclamation and reforestation, women's rights or vernacular

Figure 17 Auroville, terrace of the Visitor Centre

architecture; in others they are living a life of expatriate retirement at a standard higher than they could afford in the affluent world. Auroville is also visited by Indian tourists doing the round of temples who include Matrimandir in their itinerary.

Auroville grew out of an ashram founded by Sri Aurobindo (1872–1950). Aurobindo studied at Cambridge but on returning to India immersed himself in Vedic culture and, like Gandhi, the campaign for national liberation. Arrested for sedition in 1908, he spent his time in jail studying the Bhagavad Gita. In 1910 he arrived at Pondicherry to develop a new yoga, called integral yoga, withdrawing to his room in 1926. In 1920 he was joined by Mirra Alfassa, known as The Mother, who undertook the day-to-day management of the ashram.

In 1954, the Mother announced a dream of a new city:

> There should be somewhere on earth a place which no nation could claim as its own, where all human beings of goodwill who have a sincere aspiration could live freely as citizens of the world and obey one single authority, that of the supreme truth; a place of peace, concord and harmony . . . a place where the needs of the spirit and the concerns for progress would take precedence over the satisfaction of desires and passions, the search for pleasure and material enjoyment. . . .
>
> In short, it would be a place where human relationships which are normally based almost exclusively on competition and strife, would be replaced by relationships of emulation in doing well, of collaboration and real brotherhood.
> (Auroville, 2000: 2)

A French architect, Roger Anger, was appointed following The Mother's decision to launch the project in 1965. In 1968 the township was inaugurated with support from the government of India and UNESCO. Anger produced a plan for large, white concrete blocks spiralling out from the green centre, describing the city's image in a projected aerial view as a galaxy. He says, 'My inspiration was Le Corbusier. He was a genius in his use of form. He did not look to the past, only to the future, and he was single-minded, bold in his conceptions. He could be called the father of the new architecture' (Auroville, 2004: 87). The four radial sectors – culture, international pavilions, industry, and housing – are part of this original plan, though they were circulated by 90 degrees so that industries, such as soap and paper making, did not damage the water table. But, as I began to indicate above, there seems a tension between the original plan (maintained by more or less annual visits by Anger) and the dispersal of housing clusters through the landscape in a seemingly laissez-faire manner; and between a need for infrastructure and management of natural resources and the local agendas of dwellers. It is unclear how a vision inspired by Le Corbusier might articulate The Mother's dream, except perhaps in a common sense of sublimity.

As well as its planning department staffed by built environment professionals, Auroville has a complicated structure of working groups whose members can draw an allowance, specialist services such as the Development Group and Economy

Group, a coordinating meta-group, and a General Meeting open to all adult residents. I was told, however, that an attendance of 300 at the General Meeting is high. This raises a question as to how consensus decision-making – key to self-organization in many intentional communities and ecovillages (see Chapters Five and Seven) – can be transposed to a settlement on larger than a village scale, in effect a small town overlaid on an existing geography of several villages that remain, as Tamil settlements, separate but in part economically dependent on Auroville. A further complication is that not all the land of the proposed city and green belt zones has been purchased, while landowners begin to realize the premium they can exact from Auroville as it seeks to fill the gaps. Leaving aside such issues, some important work has been carried out. The site was deforested in 1965, mainly due to water run-off in the two annual monsoons. Over two million trees and shrubs have been planted, many grown from seeds from temple gardens where the original forest vegetation was preserved, in what is now a lush, green environment like a European idea of a tropical paradise. But there is post-modern technology, too. The Solar Kitchen is a restaurant serving communal lunches, acting as a social centre as well, built in compacted earth bricks with a 15-metre solar cooker on the roof to provide energy for all cooking and hot water – usually 1,000 meals a day. There are efforts at wind energy in housing clusters, through Auroville is not yet self-sufficient in energy. Research takes place on sustainable, mainly earth-based building materials at the Auroville Building Centre (see Maïni, 2002). In Adventure, a recent residential community, nine young architects have developed improved uses of the palm-thatch vernacular using bamboo to underhang the thatch (which houses a diversity of wildlife), granite pillars, local pine posts and outdoor kitchens (Figure 18). Such a hut can be built for £500. One has a cold tub for the very hot summer nights.

The plan is recognized as adaptable, and probably will never be more than an intention outside the central area around Matrimandir. In this environment with its vulnerable water table, realization of the plan on the scale at first envisaged might be an environmental catastrophe in itself. The plan is likely, from the evidence of the self-organizing clusters which undertake projects in Auroville, to be quietly consigned to history (which is not to say there are not material issues of water, energy and mobility for Auroville Futures to mediate). To me, in 2004, Matrimandir was more troubling than the plan. Its form is a flattened sphere outwardly covered with gilded plates, containing an inner spherical chamber accessed by spiralling staircases, one up, one down to separate the arriving from the departing viewers. Twelve meditation rooms are sited in adjoining pods, surrounded by 12 corresponding gardens. Much of the construction was undertaken by villagers employed to speed up the task. A brochure states: 'It is open for a quick visit (with passes) between 4 p.m. and 5 p.m. daily, with a possibility to stay for meditation between 5 p.m. and 6 p.m' (Auroville, 2000: 11). It reminded me of the state architecture of the inter-war period. The brochure also states: 'The Mother selected the outer form . . . from various designs which were presented to her in 1970', and that 'The Mother gave . . . very accurate indications for the design, lay-out, purpose and use of the inner room, as she saw it in a vision' (Auroville, 2000: 10–11). The following statements are included:

Figure 18 Auroville, a house in improved vernacular style

It will be a kind of hall like the inside of a column, a tower with twelve facets, each facet represents a month of the year. Right in the centre on the floor is my symbol, and above it, four of Sri Aurobindo's symbols joined to form a square, and above that a globe of 70 centimetres diameter.

The sun should enter as a ray, without diffusion: an arrangement must be made so that the ray of sunlight can be seen. When there is no sun . . . lights will be lit which have the same effect and the same colour. Right at the bottom, under the globe, there will be a light which will be directed upwards, diffusing light into the globe. It will always be in a kind of clear half-light: day and night.

Inside there will be twelve columns, situated half-way between the centre and the walls. These will be 24 metres away from the centre. No windows. Ventilation with air-conditioners. On the floor, a carpet, everywhere except for the centre. Everything will be white. No flowers, no incense nor music. People can sit everywhere. Inside no talking. Silence. . . . Everything is symbolic.

(Auroville, 2000: 10)

This contrasts with the mix of formality and informality, and varied scale, in the Visitor Centre, built in 1998 and designed by resident architect Suhasini Iyer-Guigan in compacted earth bricks with prefabricated ferrocement doorways and overhangs to take the necessary weight, natural lighting and ventilation, and solar, wind and bio-mass energy (Auroville, 2004: 49–50). The Visitor Centre was awarded a Hassan Fathy International Award for Architecture for the Poor in 1992. I end by quoting from the information booklet of the Auroville Building Centre Earth unit:

A habitat is much wider than simply the built environment, whether a house, a hospital, a school, or any urban facility. It includes the totality of the environment – not only the physical surrounding but also the social one, our neighbourhood. It takes into account a proper management of resources and development such as drainage and wastewater management systems, water and energy supply, transportation, appropriate building technologies, renewable energy sources such as sun and wind energy, use of locally available materials and skills, etc. . . . Sustainable habitat requires a holistic approach. This needs to integrate first the human aspect, which implies a different process, where people's participation is essential for its success.

(Maïni, 2002: 3)

Case 4
Christiania, Copenhagen, Denmark

At the time of writing, Christiania is under threat. News reports in March 2007 covered the eviction of squatters from the Youth House, a well-known hub of alternative living in Copenhagen, and the demolition of the building after its purchase by an evangelical Christian group. To evict the 850 or so people who live permanently in the old military buildings and self-build housing of Christiania's 85-acre riverside site would be more difficult, sending violent images to the world's news screens of a settlement which is among the most visited tourist attractions in Copenhagen. Nonetheless, Christiania's status, though established in 1971, is semi-legal. Negotiations are not helped, from the authorities' viewpoint, by the existence of several different groups with their own policies and tactics within Christiania. Police patrols remind residents they are under scrutiny, though when I went to Christiania in 2004, several police women and men were among the audience of a free rock concert. One person called them pigs but they did not react.

Walking into Christiania from a riverside path was idyllic. I saw herons in the reeds, and wandered past colourful self-build houses among trees, gardens filled with flowers, and play areas in bright sunlight. Away from the concert there were not many people, and those I passed were simply minding their own affairs. I had lunch in a small café, and from time to time noticed the smell of herbal tobacco in the air (and saw notices warning against the use of hard drugs).

The Free Town of Christiania began in September 1971 when groups of mainly young people took over redundant nineteenth-century buildings in the Christians-haven barracks, close to the city centre, in protest against the city's shortage of available housing. Jacob Ludvigsen, publisher of the youth newspaper Hovedbladet, wrote:

> Christiania is the land of the settlers. It is the so far biggest opportunity to build up a society from scratch – while nevertheless still incorporating the remaining constructions. Own electricity plant, a bath-house, a giant athletics building, where all the seekers of peace could have their grand meditation – and yogacenter. Halls where theater groups can feel at home. Buildings for the stoners who are too paranoid and weak to participate in the race . . . Yes, for those who feel the beating of the pioneer heart there can be no doubt as to the

Figure 19 Christiania, self-build, low-energy houses by the river

purpose of Christiania. It is the part of the city which has been kept secret to us – but no more.

(www.en.wikipedia.org/wiki/Freetown_Christiania)

Ludvigsen invited people looking for a home to 'immigrate with the bus no. 8 – the direct route to Christiania' (Lauritsen, 2002: 14). The squatters converted the warehouses, factories, barracks, officers' quarters and stables to new uses, and in later years began to build houses in the green spaces between the barracks and the river shore, with its remains of fortifications built to defend Copenhagen against Sweden in the seventeenth century. Coincidentally, the squatters adopted a norm of urban regeneration in transforming redundant buildings into arts spaces, with artists' studios, theatre and music performance and rehearsal spaces, galleries, and spaces for yoga and meditation. Christiania has a diversity of such spaces today, with small shops selling organic food, plants and garden items, and clothes, and a Buddhist temple, and remains car-free.

Pernille Lauritsen contextualizes the squat in terms of the aftermath of May 1968 in Paris, and a student occupation at Copenhagen University in 1970. He writes: 'people from different environment – hippies, feminists, artists, musicians, and political activists – came together in new youth and peace movements and actions' (Lauritsen, 2002: 11). A charter was drafted – still the Free Town's key document – and a weekly Common Meeting instituted to discuss matters of common concern. For Lauritsen these meetings don't work: some people shout, others are silent, and some leave because nothing is resolved (Lauritsen, 2002: 15), but perhaps this is unavoidable in learning how to construct the social architecture of a free city from scratch. The charter states:

> Christiania's objective is to create a self-governing society where each and every individual sees themselves as responsible for the well-being of the entire community. Our society is to be economically self-sustaining and our aspiration is to be steadfast in our conviction that psychological and physical destitution can be averted.
>
> (illustrated, Lauritsen, 2002: 15)

The content of this document is reflected in a short description of Christiania in a leaflet produced in 2004:

> Christiania calls itself a 'free city'. That means a sanctuary for people who do not fit in with the 'normal' rest of society, or for people who just feel a need to experience that a society can be organized differently from the rest of Denmark. . . . The houses of Christiania are a big tourist attraction. A lot of them are architecturally very special, and it is a unique experience to take a walk around the area. . . . [Christiania] is officially in favour of cannabis and has therefore allowed open trade. In return they have effectively kept hard drugs out of the area. . . . The cultural life is blossoming.
>
> (Christiania, 2004: n.p.)

After the occupation an agreement was reached with the Social Democrat government and police that the squatters could stay, and pay set rates for common water, electricity and sewage treatment. In part, funds are gained from selling Christiania's own currency to tourists (who keep the coins as souvenirs, donating 50 Kroner for each one). Numbers increased through the recession years to around 800 by 1976. Ria Bjerre lists the following: 'Zen Buddhists, Taoists, Marxists, back-to-nature folks with goats and pigs, actors from the underground theatre movement, anarchists, artists, poets, painters, musicians, filmmakers, craftspeople, academics, provincials, foreigners, drug addicts, dealers, criminals, runaways, healers, alcoholics, and yogis' (Bjerre, 1991: 88). Following this growth and the agreement with the authorities, the Free Town was (and remains) organized as 11 autonomous zones, all participating in the Common Meeting but developing their own tactics. Bjerre writes: 'There are no rules for the economic infrastructure of the free city. Those who make profits . . . are encouraged to share them, but how much they share is always up to them. In the free city nobody can force others to do what they are unwilling to do' (Bjerre, 1991: 98).

The attempt to stifle Christiania began in 1976, with a threatened evacuation (Bjerre, 1991: 8–89) followed by court cases, plans to upgrade the zone and appeals for popular support. A rainbow army of non-violent workers taking on different tasks and identified by one colour of the rainbow for each group began to improve the site, create publicity material, and galvanize support. In the end, several thousand people came to Christiania, making an evacuation too difficult. As Bjerre says, the publicity led many Danes to see Christiania as 'a good, old-fashioned European village' (Bjerre, 1991: 92), while the cost of rehousing the 800 or so residents was put at 45 million Kroner (Lauritsen, 2002: 27). Publicity led to an increase in tourism, and with it an increase in cannabis dealing – in the famous Pushers Street – that may have clouded the issue, though for many dwellers it is a non-negotiable position that cannabis use is free in Christiania. The idea of redevelopment seems, still, to have become implanted in a new proposal for a regional plan. This is ironic inasmuch as Copenhagen is known for progressive urban planning (and Denmark for tolerance and the first ecovillages). Dominic Stead writes of the 1948 Regional Plan for Copenhagen as setting out to achieve 'a pattern of decentralized concentration' with small suburbs developed as 'relatively self-contained communities' with road and rail links, limiting urban expansion to a series of corridors; this was extended in subsequent plans to 'reduce the environmental impacts of development, both in terms of resource consumption and environmental pollution' (Stead, 2000: 40–1). This is conventional planning under the power of urban and regional governance, based on planning and environmental expertise. Christiania is a model of autonomy based on the plural non-hierarchic knowledges of dwellers. Yet the use of recycled materials in the wooden houses and their innovative low-energy design might make them informative for other forms of urbanization. Pauline von Bonsdorff writes that while a built environment embodies a society's values, materials such as wood also show traces of natural ageing and 'are receptive and interactive in relation to the climate', thus avoiding the alienating and out-of-place aspect of mass housing schemes (Bonsdorff, 2005: 84–5).

In the spring of 2004, the right-wing Danish government introduced law L205 to normalize Christiania: 'If the government has its way, the first step is that everybody . . . register their personal right of property . . . Thereby the individual Christianite will get the possibility of buying their own apartment/house . . . This is a fundamental clash with the collective right of property that exists in Christiania' (Christiania, 2004: n.p.). As well as the collision between an ideology of individual property ownership and a belief in collectivity, there is the problem of drug use. Christiania has undertaken valuable work to rehabilitate hard drug users, but in 2002 a meeting of Scandinavian justice ministers in Stockholm identified it as 'the centre of the Scandinavian drug trade' (Lauritsen, 2002: 36). Since then the stalls at Pushers Street have been dismantled voluntarily. I do not know if the drug issue is a pretext for government action to close the site, but there is division within the groups in Christiania on whether the use of cannabis is integral to the free lifestyle or must now be negotiable if the autonomy of the Free Town is to survive. The difficulty is illustrated by the following statement from the *Defend Christiania* leaflet (2004):

> L205 also includes an alternative solution – the so-called 'fondsmodel' (fund model) . . . Christiania must choose a board to represent the city quarter in economic and legal matters. This solution does not immediately seem as bad . . . but how the final write up of the proposal will look is still very unclear. One could fear that the Christianites who will be on the committee will gain more power than the rest. Through this, it will be difficult to stick to the collective leadership what is central to Christiania. The goal . . . is to develop Christiania on the premisses of people who live there.
>
> (Christiania, 2004: n.p.)

The end game will probably not be about drug control or autonomy but about the development by private sector companies of highly attractive waterside sites of the kind that have featured in so many urban post-industrial speculative development schemes. In January 2006, the Danish government proposed developing new apartments for 400 people, and increases to nearer (but below) market level in the rates paid by dwellers for services. An architectural competition was organized but attracted few entries. All were rejected though the prize money of 850,000 Kroner was paid (www.en.wikipedia.org/wiki/Freetown_Christiania; see also Wheeler, 2007). I recall the delight of walking through Christiania, and the ordinariness of its domestic life. On 5 October 2004 there was a parade (Figure 20) with tractors festooned with flowers, and people in costume, some carrying banners and flags, walking to the city centre to seek poplar support for 'multiplicity, community and solidarity' (Christiania, 2004: n.p.). As the parade passed a baroque church, a red banner was unfolded from the spire.

Figure 20 Christiania, preparation for a parade

Case 5
Ufa-Fabrik, Berlin, Germany

Ufa-Fabrik occupies the redundant 1920s film processing plant of the Universal Film Studio in Berlin. It was here that German Expressionist films such F. W. Murnau's *The Cabinet of Doctor Caligari* and Fritz Lang's *Metropolis* were produced. Ufa-Fabrik was founded in 1979, and now houses the theatre-performance group International Kulturcentrum ufa-Fabrik, with two year-round stages, one of them under a circus-like tent and used as an open-air space in summer. The site has a large number of mainly low-rise buildings used as workshops, rehearsal spaces, an artists' guest house, an organic bakery producing 3,000 loaves each week as well as pastries, a café-restaurant with a summer garden, an alternative circus space and children's circus school (in which the main animals are two dogs, and most of the performance is by humans), a children's farm, an organic shop (also selling organic cosmetics, books, cards and other items), a martial arts training

Figure 21 Ufa-Fabrik, the organic café and bakery

space, a neighbourhood centre offering sport and leisure classes for people of all ages, conference venues, and a studio for the Terra Brasilis samba band (who played at the dismantling of the Berlin Wall in 1989). There is a triple pool with reed beds for grey water treatment, a pair of wind turbines on the highest building, and solar-voltaic panels on several roof spaces that generate more than 50 kilowatts of energy. Solar energy supplies Solar-Creperie, an organic crepe café. These technologies are explained on a series of noticeboards accessible to visitors. Since its foundation, the project has also housed Berlin's only certified free primary school (an offshoot of the Free University movement of the late 1960s). It also provides space for Labyrinth, a young artists' group.

The project is collectively managed by a group of 50 or so people who live and work on the site, mostly in cultural projects. The group is constituted as a commune, taking decisions by consensus in meetings open to all residents. Ufa-Fabrik's entry in *Eurotopia* (2000) reads: 'Visitors come from the neighbourhood, from the entire city and from far away . . .' (*Eurotopia*, 2000: 163), and the projects explanatory leaflet describes it as 'The artist's village in the middle of Europe's fastest changing metropolis'. Ufa-Fabrik is a member of Trans Europe Halles Network, an association of arts spaces and cultural organizations housed in redundant industrial buildings.

Ufa-Fabrik is one of a hundred or so alternative settlements in Germany, but unusual in its setting, a few U-Bahn stops from the centre of Berlin, adjacent to quiet, turn-of-the-century residential streets. While many intentional communities and ecovillages are, for either ideological or pragmatic reasons, situated in rural areas, Ufa-Fabrik shows that urban ecotopias are viable. It shows also that cultural programmes have a role in the integration of radical and ecological settlements in urban societies, since most visitors to the site come either for cultural events or educational workshops, to the café, or to buy organic products. The multi-arts events programme includes world music, tribal art, circus, cabaret, theatre and experimental forms, all with international guest appearances, attracting 400,000 visitors a year. It begins to sound much like any other cultural space, in a city that has a large resident population of artists and performers, has provided publicly subsidized live-in studio spaces for artists when property prices and rents have soared in various districts since German reunification in 1989, and is established as a cultural tourism destination. Perhaps for many of those who attend performances that is what it is, though signs advertising as organic food and renewable energy link the site to the incipient alternative society.

Ufa-Fabrik was, however, founded as a squat in 1979 when Universal moved out of the site, which was scheduled for demolition rather than re-use. The project's leaflet states: 'more than 100 people peacefully took over the desolate grounds and created a comprehensive work and living project, where daily life, profession, culture and ecology were united in a new lifestyle'. This places Ufa-Fabrik in the same tradition as Christiania (Case Four). As at Christiania, but on a much smaller scale, there is a mix of cultural and ecological projects, and a profusion of colourful decoration around the site. But while the future of Christiania is under renegotiation, and issues such as the use of cannabis are combined with those of self-organization and the evolution of consensus in a large community of diverse interests and 11 internal groupings, Ufa-Fabrik appears non-confrontational and well integrated within the city, as a visitor cultural attraction and a neighbourhood educational resource. One measure of its integration is that it employs more than 100 people.

In terms of the categories of principled and purposive, or public and personal environmentalism (Chapter Six), the cultural identity of the project suggests it is public and principled. That is, through cultural means it introduces a multicultural as well as multi-art form programme by which awareness of cultural identities in the non-affluent as well as affluent worlds gains a wider public. From that, and the content of performances that reflect the concerns of non-Western cultures and radical groups – samba, for instance, has a specific history associated with radical campaigns – other kinds of social awareness may be seeded. Of course, much of the time, for many visitors a performance is entertainment not political information. But cultural identity has become a key element in mainstream efforts to achieve greater equity, as stated in the United Nations report on cultural diversity of 1996. Having said that, my impression walking around Ufa-Fabrik on a quiet weekday morning, sitting in the café and buying a few things in the shop, was of a community getting on with its own affairs, engaged in ordinary, everyday life, hanging out but also quite productive and collectively running a successful cultural business. I wonder if the division of public and private categories is artificial in relation to such a case. Fifty or so people live there, in what amounts to an urban ecovillage where one of the key, ongoing concerns is the self-organized management of the site. To a visitor this is invisible, and photographs of people in meetings indicate little if anything of the content of discussion or the importance meetings have in consensus

Figure 22 Ufa-Fabrik, water treatment pools

decision-making, but this is possibly the utopianism of the project, derived from its beginning as a squat. Similarly, a dualism of purposive and principled is limited in explaining how everyday life in a cultural centre, which is also an urban village with its schools and neighbourhood links, either seeks to change agendas or to live out a new agenda. It appears to be doing both, the former tacitly and by the example of the latter. So I drank my coffee and ate my pastry, as a typical middle-European at mid-morning. I relaxed in the ambience of a site which demonstrates that another world is possible – not only urban dystopia – in a greening and a socializing of urban space.

Case 6
Uzupio, Vilnius, Lithuania

Uzupio is the arts district of Vilnius, the capital of Lithuania. In 1998, in protest at rising property prices in the old city that drove artists to seek less expensive studios here, across the Vilnia river, artists proclaimed the district's autonomy, not only as an arts district but as a quasi-state within the state (though without legal status). Its constitution ends with the admonition not to conquer, not to defend and not to surrender. Since 1998, ironically, Uzupio has undergone the change encountered in many other cultural quarters around the world: a first phase of cultural occupation in which artists find low-rent live-in studios has given way to a second phase of galleries and boutiques, rising rents and land prices, and the onset of a third phase of gentrification in which high property values are consolidated through renovation and a move from multiple rental to single owner occupancy.

Figure 23 Uzupio, road sign at the bridge

Uzupio feels almost like an island when approached from the old city. There is a bar with waterside tables. Further along is part of the Academy of Fine Arts. Gardens with sunflowers are visible behind the eighteenth- and nineteenth-century terraces, and the galleries have fashionably low-key entrances. In the square is a tall column bearing a sculpture of an angel, unveiled in 2002, which is now almost a logo for Uzupio. This is a fashionable district in which to live for entrepreneurs and politicians as well as arts and media professionals. It has the feel of an old city quarter, can be covered easily on foot, and still has many houses ready for renovation. From alleys behind the buildings which run uphill along the road are views across the city and its wooded hills, the sort of place in which a property boom could not fail. Its autonomy is celebrated annually on 1 April. This is in the context of a country which became self-governing in 1991 after the disintegration of the Soviet Union and a period, since liberation from the Nazis in 1945, regarded as one of foreign occupation when the Soviet liberators did not go home.

The autonomy of Uzupio trades on two sources, the history of avant-gardes in European Modernism and the vogue for cultural districts. The second is the more recent: Sharon Zukin writes that 'Every well-designed downtown has a mixed-use shopping centre and a nearby artists' quarter' (Zukin, 1995: 22), while the expansion of cultural venues such as art museums makes them 'no longer . . . adjuncts to property development . . . [but] development projects in their own right' (Zukin, 1995: 120). This is now a familiar story, replicated in city after city in competition for inward investment and cultural or business tourism. Many of the public spaces produced are in fact privatized spaces, or simulations of historical spaces (Boyer, 2001: 50). Often the reinvented cultural quarter, or the new financial district, is produced through the erasure of real histories of occupation, as in London's Docklands where images of a sparkling river complement those of the glassy towers of late Modernist architecture (Bird, 1993). But there are other possibilities. As a visitor for a few days, staying in a hotel in the old city, I can only speculate, but perhaps something of the informal economy of pre-gentrification survives in Uzupio. Looking at the cultural quarter of Nottingham's Lace Market, Jim Shorthose writes of 'strong feelings of membership of an informal creative community' and flexible networks which make 'the creative community analogous to an ecology of interdependent species rather than a formal economy of structured relations of exchange' (Shorthose, 2004: 154). Shorthose cites Ivan Illich on conviviality: 'the autonomous and creative intercourse between persons . . . an individual freedom realized in personal interdependence' (Illich, 1973: 11).

References to conviviality and interdependence place cultural autonomy in a radical tradition; and, returning to the first of the two points above, in a tradition of anti-institutional avant-gardism (Williams, 1989: 49–64). This began in the late nineteenth century in metropolitan cities when artists assumed responsibility for their own futures by taking over, as it were, the means of production in exhibition spaces. Examples are the Independent Salon in Paris and the Secessions of Munich, Berlin and Vienna. Rather than allow shifts in artistic form and content to be imposed by juries and establishments, groups of artists and critics began to organize themselves in successive departures from the mainstream. Raymond Williams sets these tendencies in art beside wider social claims to liberation – from property, economic controls and gender exclusion: 'what is new in the avant-garde is the aggressive dynamism and conscious affront of claims to liberation and creativity which . . . were in fact being much more widely made' (Williams, 1989: 57). Can Uzupio be seen as one of the more recent of such efforts, not in terms of an artistic movement but of the designation of part of a city as an art zone?

But there is another history in which to situate Uzupio, that of bohemianism. Today, new professional-class bohemians migrate to districts of inner-city gentrification. These are often adjacent to pockets of multiple deprivation, as in Hoxton, London. Part of the attraction is the raw edge and internal exile that it offers a class regarding itself as outside class categorization. Elizabeth Wilson writes of nineteenth-century bohemians in France and Germany, and the avant-gardes of early Modernism, that they were, 'drawn to utopian systems and extreme oppositionalism. Their goal was total transformation rather than incremental political change', and that their merging of art and life in an overtly alternative lifestyle (culture in the anthropological sense) expressed this (Wilson, 2003: 241). But she emphasizes that this refusal of the dominant society is not abolished or abandoned in postmodernity, when the line between high art and mass entertainment is no longer policed. If, as she writes, the shock of the new in art is no longer at all shocking, and aesthetic rebellion has tended to submerge political revolt, she continues, 'This does not mean that the desire to change society has disappeared . . . It does not follow that youth culture is devoid of political aspirations' (Wilson, 2003: 247). On the other hand, she writes also that if a period of 'democratic decay' produces a renewed possibility for an intelligentsia, still, 'when rifling through the rails of second-hand cultural identities, we find the bohemian role still hanging at the back of the wardrobe, we may feel that this old-fashioned frock is now too quaint to wear' (Wilson, 2003: 246).

One thing that has changed is the number of people belonging to a self-selecting art world. Richard Lloyd cites figures for the United States, showing the artist population of 1900 as 267 per 1,000 of the general population, and for 1999 as 900 – a more than threefold rise (Lloyd, 2006: 66, table 3.1). The spread of art education and growth of a cultural industries bureaucracy are no doubt the key enabling factors here. They have produced, together with the development of post-industrial economies in media and communications sectors, the new bohemian class seeking to live in reclaimed inner-city districts and urban villages. In the context of Vilnius, it appears Uzupio meets the need for an identifiable district, convenient to the city but also conveniently separated from it. There is a feeling of the countryside in the city in the use of the riverbank and proximity to a large riverside park and wooded hills beyond (which lead by footpaths and a scenic route to the monumental centre around the cathedral). As Lloyd writes of traditional bohemian districts, they are 'nested communities, embedded in initially poor or working-class neighbourhoods where the bohemian participants are a minority . . . As subcultures they are defined in relation to an increasingly diffuse parent culture' (Lloyd, 2006: 238). Uzupio again fits the pattern, nested in a hitherto run-down area, and like many other artist-led gentrifications it has now become the dominant element in

208 *Short case studies*

Figure 24 Uzupio, old houses by the river

the neighbourhood into which it moved. So the rents increase, and the scenario Zukin recounts is repeated in a post-Soviet situation as it was so many times on the other side of the border between an Eastern bloc and the West. Much of the former Eastern bloc is ripe for such cultural redevelopment, and I read that property prices in all the Baltic states cities have increased sharply in the past year. But, while this is undoubtedly problematic, not least for residual populations, the origin of the bohemian invasion is in the desire of artists as cultural producers to take over the means of their production, redirected here as a move to identifiable spaces in the city. What is utopian in this? First, the desire to realize a degree of freedom from (yet within) the wider society; and second, the possibility for conviviality, a frequently discussed concept in 1960s counter-cultural literature.

Case 7
Cambridge Cohousing, Massachusetts, USA

Before looking at the Cambridge Cohousing project (begun in 1996) I sketch a brief and selective history of the concept of cohousing, in the context of efforts in the past two to three decades to engage dwellers in the design process of housing, contributing to a type of urban regeneration that is socially as well as environmentally sustainable. Cohousing means the shared use of a property or group of properties by individuals and/or families when each retains individual living spaces while sharing facilities – often including a laundry and other utilities, a workshop and recreational spaces such as gardens and children's play areas. Shared meals are taken on a voluntary basis in a common dining room, usually on some rather than all days. Often, members of cohousing groups pool resources to buy vacant land on which to build, or existing properties for renovation. Typically, they evolve

Figure 25 Cambridge Cohousing, town houses

a vision for the project, get to know each other through meetings and collaboration in the initial stages of work, commission an architect, and either employ a contractor or undertake the building or renovation work themselves. There is no standard pattern and each group evolves its own way of dealing with issues that arise, but there are national networks of cohousing groups in the United States and Australia, for instance, from which to gain advice and share experiences. An advantage of cohousing is that built areas are smaller than for separate houses so that costs and energy use are lower, and many cohousing dwellers find the proximity of neighbours with whom they have bonded by participating in the project a distinct advantage over urban anonymity. Cohousing is not as intense a process as joining an intentional community and does not necessarily have the environmentalist principles of ecovillage living.

In many cases dwellers hold diverse outlooks on life and simply share a commitment to making a liveable, safe space to inhabit. Furthermore, while most intentional communities and ecovillages are in rural or peri-urban sites, cohousing tends to be in urban, often inner-city, neighbourhoods. Graham Meltzer writes that most cohousing is embedded in mainstream culture, and that 'Communities believe they can maintain internal social cohesion despite the potential tension between members' commitment to community life and their association with the world beyond' (Meltzer, 1999: 1). Matt Holland writes of the possibility that cohousing offers solutions both for deteriorating neighbourhoods, where residents may be able as a group to renovate a critical mass of properties, and areas undergoing gentrification, where cooperative purchase is cost-effective and retains the character of a neighbourhood. He argues, 'Gentrification may look nicer than ghettoization, but it's just as disruptive to the existing community' (Holland, 1993: 31). In Sabin, Portland, Oregon, the cohousing group to which Holland belongs individually acquired seven adjacent houses requiring renovation, removed boundary fences, added a common meeting room and playground, created a music room and garden with greenhouse and shed, and set up a tool bank. Cohousing offers spatial flexibility, as the nuclear family of two adults and two children for whom most housing was built in the twentieth century ceases to be a norm. Because residents are involved in design, a greater variety of spaces is included. and cohousing may be better able than standard housing to meet the needs of specific groups within society. Diana Ghirardo identifies abused women, homeless families, single mothers and rehabilitation projects as needing specific, non-standard spaces for non-traditional living (Ghirardo, 1996: 136).

The idea of cohousing began in Denmark, Sweden and the Netherlands in the 1970s, through the efforts of middle-class urban dwellers, in Sweden through women's groups, to make flexible, affordable homes (Ghirardo, 1996: 157; Barton, 2000: 74). Ghirardo cites the Danish examples of Tinngården (1979) and Tinngården 2 (1984) in Herfolge, and Savvaerket (1984) in Jystrup, designed by Tegnestuevo Vandkunsten and Karsten Vibild; and the Dutch example of Hilversum Meent (1977), designed by Leo de Long and Pieter Weeda. On Ivo Waldhör's cohousing design for a group in Malmö, Sweden in 1987 to 1991 she writes, 'the tenants had no previous experience with housing design, and all took advantage of the principle that they could make changes until construction was completed. This flexibility led to delays so that the occupants could achieve the housing they desired' (Ghirardo, 1996: 149, and plate 106). For volume house builders this would be a nightmare, yet it produced housing that meets individual needs within a single block, the exterior of which reflects the diversity of its interior arrangement.

In North America, cohousing schemes were established in 1991 and 1992 at Davis, California and Bainbridge Island, Washington state. Writing of the Winslow Cohousing project at Bainbridge Island, dweller Jane Trancho reflects: 'I feel better about my kids growing up here. . . . I know that they're in a place where, not only are they safer than in a single-family house situation, but they're also getting support and caring' (Winslow Cohousing, 1993: 39); and Kim Clark writes: 'I think we have developed a shared sense of what it takes to get along together. We are all willing to be a little softer, to listen to people, and to reconsider what we want in light of what other people want' (Winslow Cohousing, 1993: 41). Ghirardo cites Winslow, and describes the architect's role there as facilitating the vision arrived at by the group, and not providing 'philosophies about the notion of community' (Ghirardo, 1996: 158). Ghirardo then cites new urbanism as the recreation of the suburb as liveable space, accepting that new urbanist projects cater for upper-middle-class clients. To me, new urbanism is aimed at recreating the mythical safe (white) suburb of the 1950s. Meltzer argues that cohousing has 'many points in common with the utopian communities of the nineteenth century' (Meltzer, 1999: 2).

I visited Cambridge Cohousing in June 2004, going on to Arcosanti (Case Two). The south-facing 1.5-acre site is a 20-minute walk from the centre of Cambridge, near shops, cafés, schools, parks and public transport. Organized as a condominium and built with the collaboration of a developer known for cohousing schemes, the project consists of a mix of three-bedroom town houses, two-bedroom houses, and apartments of one, two, three and four bedrooms grouped in a central building, with a mix of sizes on each of four floors. There is a gradation of private and shared spaces, with common dining and sitting rooms in which meals can be taken three to five times a week and where music performances are organized, plus a laundry, utility room and garage as well as small apartments for visitors. Most gardens are unfenced (though some are individually cultivated for flowers or fruit and vegetables). There is a children's play area in front of the town houses, with colourful play structures. This was extensively used during my visit, as were the gardens. The large amount of outdoor space seemed to be important in enabling residents to live in what by North American standards are quite small units, and as an extension of domestic space used to meet other residents and share watching over children. Residents range from families with children to single people, in work or retired, several with creative or academic connections, reflecting the fact that Cambridge is a node of both contemporary cultural activity and academic institutions. The common room has a Steinway piano, and there are several musicians in or linked to the group. The project also has a library of donated books.

The aim was to build a mixed-income community, with equality of opportunities and care for environmental and conservation needs through recycling and energy reduction. There are no individual electricity meters, this being part of the common accounts and using bulk-rate purchase. Air conditioning was not installed on the grounds of noise and energy use (though the policy has been modified since). Units on the ground floor have access for people with mobility impairments. A degree of car sharing operates, though public transport limits the need for private cars – unusually for a North American city. A meal team produces common meals (devising the menu, shopping, cooking and washing up), working to a volunteer rota devised on a six-week cycle by the residents' committee. Some residents are vegetarian, others meat-eating, so options are always offered. In some cases produce is bought from a local organic farm through a community supported agriculture scheme (the equivalent of vegetable box schemes in the UK). Each member of the team brings her or his own china for the meal, avoiding the need for a decision as to what china should be bought. Similarly, the furnishings of the common areas were donated. While decisions are kept to a minimum, the project has several voluntary committees. There is a management board, and specific groups for design review, meals, events, outreach, art, children, libraries, guest rooms, laundry, mediation and advocacy, supplies and recycling. All committee decisions are posted on a board.

The project's vision statement emphasizes its urban setting, citing the vitality, variety and history of Cambridge, and the principles of 'quality, simplicity, and beauty'; it continues:

> We share a commitment to the idea that cooperating in the endeavours of daily life brings the pleasures of sociability, greater economy of resources and effort in daily tasks, the warmth of an extended family and the probability of a rich variety of friendships. In our interactions, we seek a balance between privacy in our own homes and our wish to be with others, living independently as well as interdependently. We want to share and interact with each other through social activities, celebrations and practical tasks . . . and through other shared work and problem solving. Honouring our varied experiences, we intend to follow a consensus-based process respectful of all points of view.
>
> (www.cambridgecohousing.org/about/vision.html)

Members of the initial group began in 1996 by raising capital (as a limited liability partnership) to buy the site, then a redundant industrial wasteground within a residential area, which was previously an area of brick production. The first units were occupied in 1998, and the group had what one member described as moderate control over the design. New members have since joined after belonging to a Friends group or visiting other cohousing schemes and intentional communities – one person went to Findhorn in Scotland, for instance – drawn by the combination of community and consensus decision-making with the project's urban location.

Figure 26 Cambridge Cohousing, the garden terace

Case 8
Ecovillage at Ithaca, New York State, USA

I went to Ithaca in March, 2005, thinking the early spring weather would be pleasant and mild. My geographical and climatic knowledge is not wonderful and the temperature as I arrived (by a non-ecological commuter flight from Philadelphia) was –10°C, with snow thick on the ground. After some contemplation, a taxi driver (actually the only taxi driver at the airport) agreed to take me to the ecovillage two miles from the city centre. He knew where it was, that a green store in the city was linked to it, and a few other facts about it. He seemed neither sympathetic nor unsympathetic, just driving. He dropped me where the drive into the ecovillage ran out, and I picked my way through the snow to the common house, where the day's cooking team were beginning to prepare the evening meal and somewhat bemused by my arrival. I stayed for four days in the village guest house, eating in the common dining room.

Figure 27 Ecovillage at Ithaca, the common house seen from the second phase

The site is in woodland, near the Finger Lakes that are a major tourist attraction later in the year, on a hill above Ithaca. As in many parts of New York state, most houses here have considerable land around them, and probably two cars in the drive. Ithaca is an unusual North American city, however, as the site of Cornell University, and with a significant alternative economy mainly in organic food and radical books as well as its university economy. The ecovillage describes itself as 'an intentional community and a non-profit educational organization . . . [aiming] to develop an alternative model for suburban living which provides a satisfying, healthy, socially rich lifestyle, while minimizing ecological impacts' (www.ecovillage.ithaca.ny.us). There are around 60 houses, in two phases of building, constituted as cohousing schemes, and an organic community vegetable farm with barn, greenhouses, polytunnels and a range of outbuildings, spaces for small-scale local industries, an education office, a neighbourhood root cellar built with earth-bag walls, a sheep pasture, a warm-season grasses ecosystem restoration project, and the beginning of a berry farm. The ecovillage owns the 176 acres of land on which it is built, and has acquired an adjacent plot to prevent intrusive development from taking place there. There is a large, two-storey-plus basement common house, in which common meals are taken three or four times a week, and facilities such as laundries and indoor play areas are provided as well as a sitting room for leisure and residents' meetings. A number of residents work at or have others associations with Cornell University. Some others are home-workers – alternative therapists or IT specialists, for the most part – or have their own businesses. Among future projects are the construction of an education centre, use of wind-energy, an organic orchard, biological waste water treatment, possibly cultivation of crops for bio-fuels and construction of affordable housing.

Initial land and building costs were high as a result of opting for both a site near the city, and a high standard of design, with variations on a set of modules, and material construction. Most of the houses have three bedrooms, though they range from one to four. Houses have triple glazing, passive solar energy, and six- to seven-inch dense-pack cellulose insulation made from recycled newspapers. Plumbing and wiring are run in cavities between the houses in a double-walling system. There is a common gas boiler to provide hot water and additional heating for all houses. Some houses in the second phase (SONG) use experimental green building technologies, in one case adobe-plastered wooden cavity walls filled with rammed straw bale, with a compost toilet and a Finnish tiled stove. I went inside this house, which was warm despite an outside temperature that had fallen to −15°C that evening, with only the stove providing heat from logs gathered on the site. Such high levels of insulation reduce energy use and costs by up to two-thirds of the norm in this part of New York state. Using common facilities also decreases the necessary land taken up by houses. A few residents would like, in a future phase of the site's development, to see some houses in more isolated positions in the woods where they could live entirely off the grid and self-sufficiently. Others prefer the centring of the village around pedestrian spaces, where children can play safely and there is usually someone else around to talk to.

A bus shelter has recently been constructed (in 2006), and it was part of the original planning of the ecovillage that it should be on a regular bus route. As it happens I took the bus when I left, phoning an inquiry line first to check the schedule. But I spent a lot of time walking in the four days I spent at Ithaca, partly as a defence against cold while wanting to enjoy the landscape, and also one morning as a way to have a long conversation with one of the ecovillage dwellers. I did not like to mention it to anyone, but all the time I was walking around or taking photographs of the building exteriors – including on one bright, sunny day – I noticed cars driving in or out of the site every 10 or 15 minutes. For 60 households, and about 100 adults and 63 children, it seemed a lot.

From the edge of the site or in the woods, which are being nurtured back to health after years of neglect as wasteland, with No Hunting signs all around, the ecovillage appears an isolated close-packed cluster of buildings, like a hamlet on a hill. The snow emphasized this, contrasting visually with the warm browns of the wooden houses. But the closeness of houses as a means to decrease energy use, terraced houses being more heat-retaining than detached houses, has caused difficulties for some residents, particularly since houses in the first neighbourhood (called FRoG – First Residents' Group) face each other across a narrow, winding pathway between small gardens. The view from front windows is thus of a neighbour's house. It is a pedestrian-friendly environment like a Dutch home-zone, the cars being parked at the edge of the site in a large barn-like structure. But the cars are there. Eventually a resident brought up this issue in conversation, asking if I had noticed that despite the bus stop and the intention to have access to public transport, everyone still used their cars. His explanation was that in a situation in which people are still gaining confidence in living in proximity – which he saw as an issue felt emotionally in a way different from the way it was understood intellectually, and viewed as separate from community participation in design or low-impact living – the car was a private space out of sight and hearing of others, a temporary escape from community that made a greater engagement in community possible for those for whom it was a new experience.

The extent of that engagement varies, some residents spending almost all their time in the ecovillage, others having regular employment outside. Deborah Kleiner quotes Liz Walker, the project's director, stating that, 'the biggest single challenge to making this into a mainstream movement is to streamline the process of creating community. We need a hybrid model between the way co-housing groups are currently organized and the way that standard development takes place' (Kleiner, 2000: 274). Decisions are made by consensus in open meetings or in some cases in working groups. When I was there, and sat in on two meetings, one open and the other a working group, discussion was robust but respectful. Consensus is still time-consuming, and some residents are more involved than others, particularly in the daytime when they are earning a living. Having said that, there was a sense of openness, and some continuing tension – the latter mainly over how much the ecovillage should expand, whether to three, four or five neighbourhoods. One view was that the initial plan, for five neighbourhoods to be built over 10 to 15 years, which had been agreed by consensus in the original group, should be followed. Another view was that learning from experience required the plan to be revised, and that the existing two phases were already developing separate identities, the second group suggesting they needed a common house of their own. This was interesting because there did not seem to be undue pressure on the common house, which is situated symbolically at the fulcrum of the site between the two neighbourhoods – though the plan did provide for separate common houses in each neighbourhood. Similarly, the initial aim to house 500 people of all ages seemed unreasonable to some residents, who found it enough to be in proximity to 150 (including children). One evening a complaint was voiced after the common evening meal that children playing in the basement were too noisy. Another factor was that some of those who were heavily involved in establishing the ecovillage from 1990 onwards were fatigued with organization, particularly given the frequency of meetings necessary when a community makes all decisions affecting it in common. Many of those who were involved were already escaping from the stress of the city and wider world of consumption, seeking space for personal growth and perhaps attracted to the site less because it was near the city than because, from the village amid the trees, the city was not visible.

There were a few issues apart from future development on which residents differed, such as whether pets should have free access to the woods which are also a bird habitat, or whether it should be wilderness; and whether food in the common house should be vegetarian or vegan, or include organic meat (usually there was a choice). At the time of my visit, core values were under discussion in a village re-visioning process. A majority view seemed to be that three neighbourhoods would be enough, and that the eco-footprint should be further reduced. It seemed a choice between a deepening of an ecological lifestyle for those there or a broadening to include more individuals and families, despite the considerable organizational work involved. There was, I think, a recognition that deep change at an individual level takes a long time – as in the reason given for car use – and that to go too fast might be damaging. There was also recognition that climate change will be an increasingly important issue. Perhaps the most difficult issue was whether, as one person put it, the ecovillage was a white, middle-class enclave on a hill, or a demonstration of possibility for a community of diverse class and race backgrounds. This person did not see Ithaca ecovillage as a model to be mapped onto other sites, arguing that each group in an incipient alternative society needs to find its own path and think through its own values in its own circumstances.

A majority of residents were from outside New York state. Several were from California and had participated in the Global Walk for a Liveable World in 1990, going from west to east coasts via 200 intentional communities. In June 1991 a one-week retreat in Ithaca produced the basis for an ecovillage project, under the auspices of the Center for Religion, Ethics and Social Policy, a non-profit organization linked to Cornell University. Since then several educational workshops, conferences and smaller group meetings have developed the project and devised its policies, procedures and guidelines. By 1996, when the first residents moved

Figure 28 Ecovillage at Ithaca, a street in the first phase

into houses at the site, a great deal of time and energy had been used. Whether ecovillages can become mainstream is an interesting question in the light of this history of a self-selecting community whose first members had close bonding experiences. I am not sure whether it is a matter of changing planning and design processes as much as of an incremental process of learning community, so to speak, when the mainstream culture is more concerned with privacy.

Case 9
ZEGG (Centre for Experimental Culture Design), Belzig, Germany

ZEGG (Zentrum für Experimentelle GesellschaftsGestaltung – Centre for Experimental Culture Design) is on the edge of the small town of Belzig, an hour by rail from Berlin. It occupies a site previously used by the German Democratic Republic as an espionage training school, from which several high-ranking spies graduated in the 1960s. In the 1930s it was a Nazi youth camp, the land – which includes a forest area – having been donated to the Nazis by its aristocratic owner. ZEGG was founded in 1991 with the purchase of the site by a group seeking to establish a socio-ecological settlement. Most members were from the former West Germany. Money was raised from a substantial loan which is gradually being repaid. After German reunification in 1989 land in the East was less expensive than in the West, and perhaps the site's history did not increase its value. But the relation between the Federal Republic and the Democratic Republic remains an issue, both locally where land has been bought by incomers, and for the group itself in that the Wall was a factor in some people's personal histories and remains a factor in their outlooks. There is also a certain regret that in the new conditions of a global market economy some of the benefits of the previous system in the Democratic Republic have lapsed, particularly in childcare and provision for the elderly. A member of the community at ZEGG mentioned conversations with people in the neighbourhood who recalled the informal economy in the pre-1989 period, when people shared skills. This was not a black economy in that no money was involved, but a non-money economy in which people helped each other on a basis that was understood to be generally reciprocal over time.

ZEGG is currently home to between 80 and 90 people of all ages, from single people to families with children. There is a more recent, satellite community on a nearby farm. Everyone rents their living space, from a single room to a shared flat or one-storey house. Some residents live in caravans or self-built structures in the woods. All members of ZEGG contribute to its running costs and loan repayment. In some cases people have external incomes, or are home-workers in fields such as art, craft, music, writing, film and communications. Incomes are not pooled, though this is a subject for discussion. The arts have a strong presence, and ZEGG has its own choir. A large permaculture garden provides most of the food eaten at ZEGG in communal meals. But a decision was taken not to be entirely self-sufficient in order to have a way to integrate into the local economy. ZEGG therefore buys some

ZEGG, Belzig, Germany 221

Figure 29 ZEGG, pile of woodchips and the converted boiler house

kinds of food such as dairy products, and sells its own surplus garden production locally. The forest setting provides a range of spaces for proximity or retreat, with huts and shrines in the woods. There is a waste treatment plant using reed-beds, and composting toilets. Heating is from a central system burning locally supplied wood chips, converted from the site's old boiler house (Figure 29). The site's existing buildings are being renovated to a sustainable standard, with insulation, double glazing and environmentally friendly materials. In summer the open-air, organically purified swimming pool, outside dining area and outdoor kitchen are used as guests mix with dwellers during weekly workshops and a summer camp. There is a stable for two horses used in work on the site, such as wood clearance, and a craft shop, a tailor, a bookshop, an internet café and a bar. In the centre of the site is a circular, outdoor meditation space. In addition, there are facilities for metalwork and carpentry. For the 15 or so children living at ZEGG there are play spaces and a kindergarten school. In 2001, a semi-transparent, air-flated meeting dome was piloted (Bang, 2005: 72). Following a visit by an ecological group from Burkino-Faso there is also a mud-and-timber round house at the edge of the forest.

The buildings are not the main aspect of ZEGG, however, many of them being those that were already on the site, and unexceptional in their external appearance. More significant are the everyday life of the community and its extensive programme of seminars, workshops, conferences and annual summer camp from May to September. For these there are single guest rooms and dormitory beds in a hostel, and space for up to 80 tents. The number of people on site more than doubles in the summer, which puts pressure on dwellers who find themselves living in a rather public experiment – in response to which there is an annual six-week period when ZEGG is closed to visitors. Yet such activities are seen as essential in developing individual and group awareness within the community, and as a contribution to a socially and culturally sustainable life in the wider world. In 2001, ZEGG hosted the International Communal Studies Association conference, and in 2004 hosted the Global Ecovillage Network (GEN) conference following a workshop on the Forum method that ZEGG has evolved for self- and group-knowledge, and for conflict resolution. Achim Ecker, a member of ZEGG, writes that isolated individuals or groups 'cannot create the conditions needed for a healing relationship with the Earth and its inhabitants', and that ZEGG therefore works within GEN and in the ATTAC counter-globalization movement, as well as participating in anti-nuclear demonstrations; he adds: 'to be politically effective, ZEGG needs as much communication and co-operation as possible with other active people, institutions and communities' (Ecker, 2004: 30). To enable this there is a Political Salon, to coordinate links to political groups, coalitions and new social movements.

There is also Forum, a tool for conflict resolution that can be shared and used for or passed on to other groups. Jan Martin Bang writes that Forum 'is a tool for creating sociability . . . this is not a technique readily taught . . . but requires a facilitator and an experienced group' (Bang, 2005: 70). At the pre-GEN conference workshop in 2004 a description of Forum was circulated, from which I quote at length for the sake of accuracy:

Forum is an artistic way of sharing, a stage for whatever is happening inside ourselves. . . . This focus on transparency, sharing and clarifying unsolved situations of daily life makes it an invaluable catalyst for one's own growth. . . . A meaningful Forum needs a mental-spiritual basis among the group utilizing it. Forum is designed to work with people who are living together, sharing a common vision and who are committed to certain values such as trust, love, solidarity and responsibility. A central and essential value for Forum is trust. . . .

What comes to the surface when we begin working in Forum is not always nice. In the beginning, the suppressed and hidden emerge into the light of awareness. However, an effective and skilful Forum will bring out the dark side with humour, or in some theatrical way so that it can be perceived without judgement. Forum wants to lift the energy level, wants to trigger the life force and its expression. When the energy can be successfully raised a change of perspective on both the body and soul level happens. Sometimes this energy shift can be very simple, as when the facilitator invites the presenter to move faster, or to exaggerate gestures, or to put a sound to the feeling. Trying out different ways of behaviour and theatrically acting out emotional processes is an important step towards dis-identification. I come to see that I am not this anger. I am not this fear. I am not this jealousy. To lose identification with these passing states means that you have found an inner position of witnessing what is going on, of standing back from it. You have found your unchanging centre. At the same time, Forum is no substitute for each individual's ongoing inner work.

The method of Forum involves sitting in a circle with a facilitator. One person enters the space in the middle and speaks, acts, sings, dances, mimes, or otherwise expresses a feeling, idea or response to a situation. In the centre they have uninterrupted attention from all present, while the facilitator may ask a question or make a suggestion. When the person returns to her or his seat, another may enter the space to describe what they have witnessed, or react. The process usually lasts for about an hour and a half, until those who feel so moved have entered the circle. Forum may emphasize personal liberation but in fact depends on group solidarity and use of performative communication, in that if a person's understanding of themselves and the perceptions of others they hold is opened, it enables others' perceptions to be made public in a situation of trust. Within a self-organizing community it is a vital means to deal with conflicts and the unease that is inevitable in living in an experimental way. For this reason, ZEGG has undertaken Forum workshops for other groups, ecovillages and intentional communities, as well as offering workshops such as that in which I participated in 2004.

The forest setting, outdoor kitchens and camps, and the prominence of performative conflict resolution at ZEGG suggest an element of the tribalism of the Dongas (Chapter Seven) or the Rainbow Gatherings (Chapter Five). This is referred to in ZEGG's own material:

> We are living in a culture of separation ... We live separated from Nature, treating it as a source of raw materials ... that's why what we understand as individuality is often so hollow and has to be proven and maintained with the help of so many external props. Contrast this with the interconnected way of life which is still practised ... by so-called indigenous peoples. There a person lives as part of her or his community, the tribe, and the tribe in turn lives as an integral part of the surrounding Nature. We're not advocating a return to such tribal cultures, but we can develop ... contemporary forms of community which reflect our cultural history. In this way we make a conscious decision to live in community.
>
> (www.zegg.de/index.php?communit)

In this context children learn experientially, at an early age in the kindergarten and play spaces, and later at a nearby Free School or local state schools. Children take part in the community's daily life, and have a house for group activities. Young people can live separately from their parents, organizing their affairs and participating in their own Forums. A particular emphasis is put on women's liberation, or woman-power as community. Leila Dregger, a member of ZEGG and editor of the magazine *Frauenstimme* (*Women's Voice*) writes, 'Some women felt responsible for intimacy, love and sexuality in the community. They tackled questions such as: where do those who don't have a partner find emotional and sexual intimacy, and security? Does a certain couple need help in their communication? Do the young people need more support or protection?' (Dregger, 2002: 84). Through sharing and drawing out feelings such as jealousy and competition, and seeing these historically in the context of patriarchy, women at ZEGG found stronger trust and fuller capacity to love without the limitations of conventional single-partner relationships. Dregger continues, 'We found that the more truth can be risked in a community, the bigger its chances of survival are' (ibid.).

Alongside permaculture and an environmentally low-impact way of life, then, ZEGG has evolved a means to build new social architectures. Part of its social architecture is the practise of free love, another its use of Forum to work on questions of love and sexuality. This is viewed as integral to creating a viable self-organizing society:

> Love is the home of the divine on earth and the quality that makes us human beings. Love and sexuality are sources of life. We see it as an essential task to create ways of living that integrate these sources in a conscious and positive way. 'Free Love' in this sense for us is an all encompassing cultural work to heal these sources of life.
>
> A peaceful culture is rooted in solidarity between the sexes. Many people see the issues of love and sexuality as exclusively personal affairs and try to solve them in private or within the framework of therapy. We see love as a political issue as social and cultural changes are needed for the development of love.
>
> (www.zegg.de/index.php?consciousnessinlove)

ZEGG, Belzig, Germany 225

Figure 30 ZEGG, the treehouse for lovers

A similar commitment is made by some groups in Christiania (Case Four). Julia Kristeva, too, writes of the sexual revolution that was part of the events leading up to the insurrection of May 1968 in Paris: 'I think it's indispendsible, because it's this liberated behaviour that contributed to the liberation of French society . . . '68 was a worldwide movement that contributed to an unprecedented reordering of private life' (Kristeva, 2002: 18). In practical terms, the difficulties of sexual and emotional relationships will tend to disrupt a small, close-living community's other work unless addressed. For ZEGG, 'If communication about love is cultivated, love can become more free of the fear of abandonment, lies and pretence . . . the more we are able to handle difficulties in a playful and unidentified way, the less we depend on this one moment of fulfilment, the less we tend to accusations, the more truth we will risk and the more power and love we have to give the world' (www.zegg.de/index.php?consciousnessinlove). Hence members of ZEGG live in a variety of relationships and sexual or other friendships, between individuals of the same or other gender. In the woods at the satellite community is a treehouse, suspended on a cable to be lowered and raised, which two people can arrange to use for a special night (Figure 30).

I found it difficult at first to participate in Forum. Each session began with a communal song – I have an inability to enjoy such acts. Gradually I found affinities. Only on the final day did I step into the circle, though I had given a short talk about this book at an informal session the previous evening. On the last night the ZEGG choir sang from Pablo Neruda's *Canto general*. In the morning, I took the bus, the train . . . back to the outside world. But I had changed: my guarded consciousness was a little re-inflected.

Conclusion

In this conclusion I offer no definition of Utopia. To do so would be prescriptive, when I have argued that if a better world is to be produced through human agency it will entail a handing over of the process at grassroots level. All I can attempt is to sum up some lines of argument and suggest some issues for further consideration.

In Part One (Chapters One to Three), I argued that literary Utopias are critical devices that distance their authors from the content of the text, reflecting on present reality rather than proposing that the society described should be realized. If Thomas More was attracted to the monastic aspect of the life described in *Utopia*, I speculated this does not mean that he sought to transform England into a monastic community, but rather to draw attention to the irrationality of its justice. I argued that a misrepresentation of literary Utopias as prescriptive texts becomes a flaw in the concept of Utopia when the power relations of the existing society are reproduced in interpretations of the world for others by an elite (or later by an avant-garde). Such interpretations-for are like power-over, implicitly assuming those others for whom the interpretation is intended cannot interpret – or change – the world for themselves. I argued also that René Descartes' search for certainty was a utopian project in context of religious wars, but that his image of an engineer drawing a line – a foundational image for modernity – is a metaphor. I then suggested that Charles Fourier's proposed reorganization of France on the basis of passional attraction, located in phalansteries, could also be read as a literary text, even while Fourier presents it as a plan (and no doubt believed in the efficacy of his intricately detailed tabulations).

In Part Two (Chapters Four to Seven) I moved to modern architecture and planning as efforts to engineer a better world by design, contrasting the pastoral vision of the Garden City with Le Corbusier's aggressive Modernism. I prefaced discussion of both by outlining Idelfons Cerdá's 1859 plan for the northern extension of Barcelona, which demonstrated a progressive urbanism centred on the means of human well-being while assuming that these could be provided in a built environment, and that such provision would limit the likelihood of insurrection. In Chapter Five, however, I related aspects of the cultural insurrection of the Summer of Love in San Francisco in 1967, drawing attention to its performative and political dimensions. This was a moment of departure from the dominant society, as emphasized in a move to intentional communities in remote places where the social architectures of an alternative society could be built through means such as consensus decision-making and low-impact living. In Chapter Seven I observed continuities from the 1960s in a renewed growth of alternative living in ecovillages, and renewed departure from the dominant society in activism. In Chapter Six, as a pivot between discussion of the 1960s and the 1990s, I noted the growth of green consumerism and a breadth of new kinds of awareness, or knowledges, now produced in the varieties of environmentalism. Among my key concerns in Part Two were the need to learn from the experiences of grassroots environmentalism in the non-affluent world, and to understand the relation of means to ends in the search for a world indisputably better than that of present socio-economic organization, based on principles of social and environmental justice.

In Part Three, then, I focused on such experiences and the means employed in projects in the non-affluent world. In Chapter Eight I found Hassan Fathy's mud-brick architecture had much to offer in terms of low-impact living, and that he saw the construction of settlements as a cooperative effort; but I cited arguments that he nonetheless retained the perspective of an urban professional, a member of an elite class able to facilitate improvements in village housing but not to hand over the means to create a housing revolution. In Chapter Nine I discussed the work of the Barefoot College in Rajasthan as exemplifying such a handing over, in the context of post-colonial frameworks of culture and education. I am reticent in claiming that the Barefoot College is a utopian society, in part because the concept of a Utopia is integral to a Western modernity based on a trajectory of progress, and it is not appropriate to map the concepts of one culture onto the alternative societies of another. Having said that, the Barefoot College departs from the Indian tradition of social work, just as the move to intentional communities in the West in the 1960s (when many young Westerners went to India) was a departure from a world of competitive capitalism. This brings me to two further issues: whether the shift of awareness that produces, and is produced by, a new social formation operates primarily at the individual or the social level; and whether a shift of consciousness, at a deep level for both individuals and the mass publics and communities of interest to which they belong, is possible (and if so how). A wall at ZEGG (Case Nine) is painted colourfully with the slogan 'Another World is Possible' (Figure 31), but how will it come into being?

There are relevant philosophical issues in the European concept of emancipation, read as a move to a radically different state of society separated from the present by an apocalyptic chasm, as Ernesto Laclau (1996) puts it, or (as he also discusses) as an incremental shift through reform rather than revolution – which resembles Raymond Williams' idea of a long revolution (1961). Laclau argues that either way the terms of the departure are set within the conditions from which the departure takes place, and that a negotiated position between freedom and unfreedom is what is possible in a democratic society today. This might be taken as a refusal of Utopia in favour of a more gradual process of change, in the case of environmentalism and anti-capitalism pursued in coalitions of local action networked on a global scale.

Conclusion 229

Figure 31 ZEGG, 'Another World is Possible'

The question as to whether a new individual consciousness is a prerequisite for social change, raised in the 1960s following contact with Eastern religions such as Buddhism, continues to surface in 1990s activism. I have argued above that such a personal moment of liberation amid others is the new awareness produced in activism and ecovillage living. The work of ZEGG on consciousness, community, and love and sexuality appears to be the research and development work, so to speak, of a new society – not as a means to an end but as the end in practice. ZEGG, however, is in some ways a departure from, or a more developed position within, a Global Ecovillage Network (Figure 32) that includes diverse viewpoints, in many cases using methods such as permaculture and renewable energy but plural in its relations to the dominant society and wider political formations. But I argued also, from the examples of the Barefoot College and Coventry Peace House, that the Gandhian principles of non-violence and working with the poorest of the poor constitute utopian tactics as much as achieving a reduced environmental footprint. These are, of course, compatible aims, not alternatives. And perhaps most such dualisms, like the relation of individual and social shifts in consciousness, can be regarded as non-oppositional polarities on an axis of potentially creative tension.

After the failure of revolt in Paris in 1968, in a period of retrenchment for the New Left in Europe and North America, Herbert Marcuse proposed a new sensibility as the vehicle for a new society. In *An Essay on Liberation,* he writes:

230 *Conclusion*

Figure 32 ZEGG, members of the Global Ecovillage Network at the 2004 annual meeting

> The new sensibility, which expresses the ascent of the life instincts over aggressiveness and guilt, would foster, on a social scale, the vital need for the abolition of injustice and misery and would shape the further evolution of the 'standard of living'. The life instincts would find rational expression (sublimation) in the planning and distribution of the socially necessary labour time within and among the various branches of production . . . the liberated consciousness would promote the development of a science and technology free to discover and realise the possibilities of things . . . playing with the potentialities of form and matter . . . the opposition between imagination and reason, higher and lower faculties, poetic and scientific thought, would be invalidated. Emergence of a new Reality Principle: under which a new sensibility and desublimated scientific intelligence would combine in the creation of an *aesthetic ethos*.
>
> (Marcuse, 1969: 31–2)

Elsewhere, Marcuse argues for the radical idea of society as a work of art, at which the last sentence above points (Marcuse, 1968a). It is an attractive idea, but appears to rest on what amounts to a biological transformation of consciousness, when the technology for an ending of the economic problem of scarcity is contradicted by an inequitable distribution of the beliefs of technological progress; this situation then prolongs or widens social divisions. It is effectively a reworking of the nineteenth-century idea of freedom as the end of scarcity in terms of biology and psychoanalysis. I have argued elsewhere (Miles, 2004a: 70–92) that one of the difficulties – articulated in Marcuse's lectures in Berlin in 1967 (Marcuse, 1970) – is that Utopia is seen both as the ending of scarcity and as a tomorrow temporally divided from present actuality. My argument was that, citing Henri Lefebvre's theory of moments of liberation within the routines of everyday life (see Shields, 1999: 58–64), a new world can be envisaged as co-present with the dominant society in such moments. This shifts the ground of the problem from time to a metaphorical space. In terms of the material of this book, I would argue that the consciousness produced in activism is a collective moment of liberation in Lefebvre's terms, experienced in personal consciousness but in a shared situation. I doubt, therefore, a need to resort to biological explanations, though, as cited in Chapter Six, I note Elizabeth Gross' reading of Darwin's theory of evolution as stating that biology, too, is produced, and thus has its own histories. I have also written elsewhere (Miles, 2004a: 147–79) on contemporary dissident art that contributes to a radical re-visioning of present society through exposure of its realities and development of collaborative means to recreate agency in ways appropriate to social or environmental justice. Here I would again mention the work of the Freee Art Collective, as in the project 'Futurology' in 2004, one part of which involved working with school students to imagine their own futures (Figure 33).

Perhaps the idea of Utopia becomes less important in the practices of an alternative society if Utopia is taken as an eventual or objectively given end of history. The point is that in a genuinely free society, history is an unending process of adaptation that differs from a Darwinian universe only in the working through of principles – such as non-violence and mutual aid – emerging in human agency, recognizing that if biology is non-teleological then history may be so as well. This adds emphasis to the idea that means enact values, and as such are ends. Perhaps in the face of the unlikely occurrence of a real change in political structures, it is necessary to look to personal awareness or aesthetic life for a sense of freedom, or for a refusal to accept the lies of the affluent society.

Marcuse writes: 'the rebellion . . . [is] against this false, illusory transfiguration of meaninglessness into something meaningful in art' (Marcuse, 2007: 126), and also writes of the art of intimacy, in love stories, as the resort of freedom under conditions of total oppression (Marcuse, 1998: 199–214). Perhaps this is the arena in which one kind of negotiation of a place between freedom and unfreedom occurs, where glimpses of a utopian realm of transformation are experienced. It occurs, too, in the empowerment that marginalized people realize for themselves in grassroots activism, and in low-impact living in the West. These are triangulation points for a utopian terrain for the twenty-first century. It is neither plan nor social engineering, nor adherence to the dreams of elites or avant-gardes, but presence in a moment in which the noise of the dominant society ceases, and where another world is possible.

232 Conclusion

Figure 33 'Mapping the Future', Dave Beech (Freee Art Collective), Futurology project, Walsall Art Gallery, 2004 (photo A. Hewitt by permission of Freee)

Bibliography

Adorno, T.W. and Horkheimer, M. (1997) *Dialectic of Enlightenment*, London, Verso [first published 1944, New York, Social Studies Association]

Affron, M. and Antliff, M., eds (1997) *Fascist Visions: Art and Ideology in France and Italy*, Princeton (NJ), Princeton University Press

Aga Khan Award for Architecture (2001) *Modernity and Community: Architecture in the Islamic World*, London, Thames and Hudson

Agarwal, B. (1998) 'The Gender and Environment Debate', in Keil *et al.* (1998) pp. 193–220

Albert, M. (2001) 'Anarchists', in Bircham and Charlton (2001) pp. 321–7

Amin, S. (2001) 'The Globalization of Social Struggles', in Houtart and Polet (2001) pp. 60–2

Antliff, M. (1997) '*La Cité Française* – Georges Valois, Le Corbusier and Fascist Theories of Urbanism', in Affron and Antliff (1997) pp. 134–70

Arendt, H. (1958) *The Human Condition*, Chicago, University of Chicago Press

Arendt, H. (1971) *The Life of the Mind* (Book I, *Thinking*; Book II, *Willing*), New York, Harcourt

Ashcroft, B., Griffiths, G. and Tiffin, H., eds (1995) *The Post-Colonial Reader*, London, Routledge

Ashton, O. (1999) 'Orators and Oratory in the Chartist Movement, 1840–1848', in Ashton *et al.* (1999) pp. 48–79

Ashton, O., Fyson, R. and Roberts, S., eds (1999) *The Chartist Legacy*, Woodbridge (Suffolk), Merlin Press

Auroville (2000) *Auroville: A Dream Takes Shape*, 7th edn, Auroville, Alain G.

Auroville (2002) *Auroville*, Auroville, AV Publication Group

Auroville (2004) *Auroville Architecture: Towards New Forms for a New Consciousness*, Auroville, Auroville Foundation

Avineri, S. and de-Shalit, A., eds (1992) *Communitarianism and Individualism*, Oxford, Oxford University Press

Bahuguna, S. (1990) 'Chipko – From Saving the Forests to the Reconstruction of Society', in Woodhouse (1990) pp. 111–21

Bang, J. M. (2005) *Ecovillages: A Practical Guide to Sustainable Communities*, Edinburgh, Floris Books

Barker, C. (2001) 'Socialists', in Bircham and Charlton (2001) pp. 329–36

Barker, F. (1984) *The Tremulous Private Body: Essays on Subjection*, London, Methuen

Barton, H., ed. (2000) *Sustainable Communities: The Potential for Eco-Neighbourhoods*, London, Earthscan

Barton, H. and Kleiner, D. (2000) 'Innovative Eco-Neighbourhood Projects', in Barton (2000) pp. 66–85

Bibliography

Barry, J. (1999) *Environment and Social Theory*, London, Routledge
Bauman, Z. (1989) *Modernity and the Holocaust*, Cambridge (UK), Polity Press
Bauman, Z. (1998) *Globalization: The Human Consequences*, Cambridge (UK), Polity Press
Bauman, Z. (2000) *Liquid Modernity*, Cambridge (UK), Polity Press
Beck, G. (1991) 'The Rainbow Gatherings', in Buenfil (1991) pp. 67–82
Beecher, J. and Bienvenu, R., eds (1983) *The Utopian Vision of Charles Fourier: Selected Texts on Work, Love and Passionate Attraction*, Columbia (MS), University of Missouri Press
Beevers, R. (1988) *The Garden City Utopia: A Critical Biography of Ebenezer Howard*, London, Macmillan
Bell, D. and Jayne, M., eds (2004) *City of Quarters: Urban Villages in the Contemporary City*, Aldershot, Ashgate
Belsey, C. (1985) *The Subject of Tragedy: Identity and Difference in Renaissance Drama*, London, Routledge
Bender, T. (1999) 'Intellectuals, Cities, and Citizenship in the United States: The 1890s and 1990s', in Holston (1999) pp. 21–41
Benjamin, W. (1997) *Charles Baudelaire: A Lyric Poet in the Era of High Capitalism*, London, Verso [first published 1973, London, New Left Books]
Benjamin, W. (1999) *The Arcades Project*, Cambridge (MA), Harvard
Benjamin, W. (2006) *On Hashish*, Cambridge (MA), Harvard
Berger, M. (1964) *The New Metropolis in the Arab World*, New Delhi, Allied Publishers
Berg-Schlosser, D. (2003) 'Conclusions and Perspectives', in Berg-Schlosser and Kersting (2003) pp. 217–22
Berg-Schlosser, D. and Kersting, N., eds (2003) *Poverty and Democracy: Self-Help and Political Participation in Third World Cities*, London, Zed Books
Bernstein, R. (1985) *Beyond Objectivism and Relativism: Science, Hermeneutics, and Praxis*, Philadelphia, Temple University Press
Berry, K. A. (1998) 'Race for Water? Native Americans, Eurocentrism, and Western Water Policy', in Camacho (1998) pp. 101–24
Bingaman, A., Sanders, L. and Zorach, R., eds (2002) *Embodied Utopias: Gender, Social Change and the Modern Metropolis*, London, Routledge
Binnie, J., Holloway, J., Millington, S. and Young, C., eds (2006) *Cosmopolitan Urbanism*, London, Routledge
Bircham, E. and Charlton, J., eds (2001) *Anti-Capitalism: A Guide to the Movement*, London, Bookmarks Publications
Bird, J. (1993) 'Dystopia on the Thames', in Bird *et al.* (1993) pp. 120–35
Bird, J., Curtis, B., Putnam, T., Robertson, G. and Tickner, L., eds (1993) *Mapping the Futures: Local Cultures, Global Change*, London, Routledge
Bjerre, R. (1991) 'For Christiania with Love', in Buenfil (1991) pp. 85–100
Black Bear Ranch (1974) *January Thaw: People at Blue Mountain Ranch Write about Living Together in the Mountains*, New York, Times Change Press
Blackman, W. S. (1927) *The Fellahin of Upper Egypt*, London, Harrap
Blackwood, C. (1984) *On the Perimeter*, London, Flamingo
Blanchard, T. (2000) 'Bending the Rules', *Observer Magazine*, 16 July, pp. 56–9
Blau, E. and Troy, N. J., eds (2002) *Architecture and Cubism*, Cambridge (MA), MIT
Bloch, E. (1986) *The Principle of Hope*, Cambridge (MA), MIT [previously published in German (1959) as *Das Prinzip Hoiffnung*, Frankfurt am Main, Suhrkamp Verlag]
Boal, A. (1979) *Theatre of the Oppressed*, London, Pluto [new edition 2000, London, Pluto Press]

Bolman, R. (2002) 'Natural Building and Social Justice', in Kennedy *et al.* (2002) pp. 14–15
Bonsdorff, P. von (2005) 'Building and the Naturally Unplanned', in Light and Smith (2005) pp. 73–91
Borden, I., Kerr, J., Rendell, J. and Pivaro, A., eds (2001) *The Unknown City: Contesting Architecture and Social Space*, Cambridge (MA), MIT
Bové, J. and Dufour, F. (2001) *The World is Not for Sale: Farmers Against Junk Food*, London, Verso
Boyer, C. (2001) 'Twice-told Stories: the Double Erasure of Times Square', in Borden *et al.* (2001) pp. 30–52
Braunstein, P. and Doyle, M. W., eds (2002) *Imagine Nation: The American Counter-Culture of the 1960s and 1970s*, New York, Routledge
Bretting, J. G. and Prindevill, D-M. (1998) 'Environmental Justice and the Role of Indigenous Women Organizing Their Communities, in Camacho (1998) pp. 141–64
Bronstein, J. L. (1999) 'From the Land of Liberty to Land Monopoly: the United States in a Chartist Context', in Ashton *et al.* (1999) pp. 147–70
Broome, J. and Richardson, B. (1995) *The Self-Build Book*, 2nd edn, Totnes (Devon), Green Books
Buber, M. (1996) *Paths to Utopia*, Syracuse (NY), Syracuse University Press [first published 1949]
Buenfil, A. R. (1991) *Rainbow Nation Without Borders: Toward an Ecotopian Millennium*, Santa Fe (NM), Bear and Company Publishing
Camacho, D. E., ed. (1998) *Environmental Injustices, Political Struggles: Race, Class, and the Environment*, Durham (NC), Duke University Press
Campanella, T. (1981) *The City of the Sun*, trans. Donno, D. J., Berkeley, University of California Press
Capra, F. (1995) 'Deep Ecology: A New Paradigm', in Sessions (1995) pp. 19–25
Carey, P., ed. (1999) *The Faber Book of Utopias*, London, Faber and Faber
Carmen, R. (1996) *Autonomous Development: Humanizing the Landscape; An Excursion into Radical Thinking and Practice*, London, Zed Books
Carney, J. A. (1996) 'Converting the Wetlands, Engendering the Environment: The Intersection of Gender with Agrarian Change in Gambia', in Peet and Watts (1996) pp. 165–87
Carpenter, K. (1984) *Desert Isles and Pirate Islands*, Frankfurt am Main, private publication
Çelik, Z. (2000) 'Le Corbusier, Orientalism, Colonialism', in Rendell *et al.* (2000) pp. 321–31
Césaire, A. (1997) *Return to My Native Land*, London, Bloodaxe Books
Chambers, N., Simmons, C. and Wackernagel, M. (2000) *Sharing Nature's Interest: Ecological Footprints as an Indicator of Sustainability*, London, Earthscan
Chambers, R. (1998) 'Us and Them: Finding a New Paradigm for Professionals in Sustainable Development', in Warbruton (1998) pp. 117–47
Chapman, E. (1998) 'Living Lightly on the Land', *Permaculture*, 17, pp. 20–2
Choay, F. (1997) *The Rule and the Model: On the Theory of Architecture and Urbanism*, Cambridge (MA), MIT
Christiania (2004) *Defend Christiania! – Forsvar Christiania!* Christiania, Copenhagen [leaflet published by Forsvar Christiania campaign, not paginated]
Cilliers, P. (1998) *Complexity and Postmodernism: Understanding Complex Systems*, London, Routledge
Claeys, G., ed. (1994) *Utopias of the British Enlightenment*, Cambridge (UK), Cambridge University Press
Clark, W. (1989) 'Managing Planet Earth', *Scientific American*, 261(3), pp. 46–57

236 Bibliography

Coates, C. (2001) *Utopia Britannica: British Utopian Experiments 1325–1945*, London, Diggers & Dreamers Publications

Cobb, J. H. Jr (1993) 'Comments on *Arcosanti: An Urban Laboratory?*', in Soleri (1993) p. 9

Colomina, B., ed. (1992) *Sexuality and Space*, New York, Princeton Architectural Press

Colomina, B. (1996) *Privacy and Publicity: Modern Architecture as Mass Media*, Cambridge (MA), MIT

Connor, S. (1997) 'The Modern Auditory I', in Porter (1997) pp. 203–23

Cook, J. (2001) 'Preface' to Soleri (2001) pp. 11–16

Coombes, A. (1994) *Reinventing Africa: Museums, Material Culture and Popular Imagination*, New Haven (CT), Yale

Cooper, D., ed. (1968) *The Dialectics of Liberation*, Harmondsworth, Penguin

Corbin, A. (1996) *The Foul and the Fragrant: Odour and the Social Imagination*, Basingstoke, Macmillan [published 1994 as *The Foul and The Fragrant: Odour and the French Social Imagination*, London, Picador; and 1986 as *The Foul and the Fragrant: The Sense of Smell and its Social Image in Modern France*, Oxford, Berg]

Cosgrove, D. (1994) 'Contested Global Visions: One-World, Whole-Earth, and the Apollo Space Photographs', *Annals of the Association of American Geographers*, 84(2), pp. 270–94

Cottingham, J. (1997) *Descartes*, London, Phoenix

Cottingham, J., Stoothoff, R. and Murdoch, D. (1985) *The Philosophical Writings of Descartes*, vols. I and II, Cambridge (UK), Cambridge University Press

Crouch, D. (2003) *The Art of Allotments: Culture and Cultivation*, Nottingham, Five Leaves

Curtis, B. (2000) 'The Heart of the City', in Hughes and Sadler (2000) pp. 52–65

Curtis, K. (1999) *Our Sense of the Real: Aesthetic Experience and Arendtian Politics*, Ithaca (NY), Cornell University Press

Damisch, H. (1995) *The Origin of Perspective*, Cambridge (MA), MIT

Darwin, C. (1996) *The Origin of Species*, Oxford, Oxford University Press [first published 1859]

Dean, A. O. and Hursley, T. (2002) *Rural Studio: Samuel Mockbee and an Architecture of Decency*, New York, Princeton Architectural Press

Degen, M. (2004) 'Barcelona's Games: The Olympics, Urban Design, and Global Tourism', in Sheller and Urry (2004) pp. 131–42

Deloria, P. (2002) 'Counterculture Indians and the New Age', in Braunstein and Doyle (2002) pp. 159–88

Desai, P. and Riddlestone, S. (2002) *Bioregional Solutions For Living on One Planet*, Totnes (Devon), Green Books

Descartes, R. (1925) *Discours de la méthode*, ed. Gilson, E., Paris, Vrin

Descartes, R. (1960) *Discourse on Method*, Harmondsworth, Penguin

Descartes, R. (1963) *Oeuvres Philosophiques*, 3 vols, Paris, Garnier

Dethier, J. (1982) *Down to Earth – Adobe Architecture: An Old Idea, a New Future*, London, Thames and Hudson

Dewey, J. (1984) *The Quest for Certainty*, in *The Later Works, 1925–1953*, vol. 4, Carbondale (IL), Southern Illinois University Press [first published 1929]

Didion, J. (2001) *Slouching Towards Bethlehem*, London, Flamingo

Diggers and Dreamers (2004) *The Guide to Communal Living 2004/2005*, London, Diggers & Dreamers Publications

Dobson, A. (1995) *Green Political Thought*, 2nd revised edn, London, Routledge

Docker, J. (1995) 'The Neocolonial Assumption in University Teaching of English', in Ashcroft *et al.* (1995) pp. 443–6

Donald, J. (1999) *Imagining the Modern City*, London, Athlone

Doughty, D. (2002) 'From the Nile to the Rio Grande', in Kennedy *et al.* (2002), pp. 249–53

Doyle, M. W. (2002) 'Staging the Revolution: Guerrilla Theatre as a Countercultural Practice, 1965–68', in Braunstein and Doyle (2002) pp. 71–98

Dregger, L. (2002) 'Women at Zegg, Germany', in Jackson and Svensson (2002) pp. 84–5

Dunster, B. (2003) *From A to ZED: Realising Zero (Fossil) Energy Developments*, London, Bill Dunster Architects/ZEDfactory

Dürrschmidt, J. (1999) 'The Local Versus the Global? Individual Milieux in a Complex Risk Society: The Case of Organic Food Box Schemes in the South West', in Hearn and Roseneil (1999) pp. 131–54

Eccleshall, R., Geoghegan, V., Jay, R., Kenny, M., MacKenzie, I. and Wilford, R. (1984) *Political Ideologies: An Introduction*, 2nd edition, London, Routledge

Ecker, A. (2004) 'Preparations for a New World: An Experimental Community in Germany', *Permaculture*, 39, pp. 27–30

Edwards, A. (1982) *A Thousand Miles Up the Nile*, London, Century Publishing [first published 1877]

Edwards, S., ed. (1970) *Selected Writings of Pierre-Joseph Proudhon*, London, Macmillan

Ekram, L. (1997) 'Aranya Low-Cost Housing, Indore, India', in Serageldin (1997) pp. 106–11

Elliott, D. (2003) *A Solar World: Climate Change and the Green Energy Revolution*, Totnes (Devon), Green Books

Elliott, J. (1999) *An Introduction to Sustainable Development*, 2nd revised edn, London, Routledge

Engels, F. (1892) *The Condition of the Working Class in England in 1844*, London, Allen and Unwin

Engels, F. (1908) *Socialism – Utopian and Scientific*, 2nd edn, Chicago, Charles H. Kerr Company [1st edn 1903]

Escobar, A. (1996) 'Constructing Nature: Elements for a Post-structural Political Ecology', in Peet and Watts (1996) pp. 46–68

Eurotopia: Directory of Intentional Communities and Ecovillages (2000) Poppau (Germany), Okodorf Sieben Linden

Fairfield, R. (1972) *Communes USA: A Personal Tour*, Baltimore (MD), Penguin

Fairlie, S. (1998) *Low Impact Living: Planning and People in a Sustainable Countryside*, East Meon (Hampshire), Permanent Publications

Fairlie, S. (2003) 'Planning for Change', *Permaculture*, 36, pp. 17–20

Fanon, F. (1967) *The Wretched of the Earth*, Harmondsworth, Penguin [first published 1961 as *Les damnées de la terre*, Paris, Masparo]

Fanon, F. (1986) *Black Skin, White Masks*, London, Pluto Press [first published 1952 as *Peau noire, masques blanc*, Paris, Editions de Seuil]

Farber, D. (2002) 'The Intoxicated State/Illegal Nation: Drugs in the Sixties Counterculture', in Braunstein and Doyle (2002) pp. 17–40

Farrell, J. J. (1997) *The Spirit of the Sixties: The Making of Post-war Radicalism*, New York, Routledge

Fathy, H. (1969) *Gourna: A Tale of Two Villages*, Cairo, Egyptian Ministry of Culture [republished 1973 as *Architecture for the Poor*, Chicago, University of Chicago Press, and 1989 as *Architecture for the Poor*, Cairo, American University Cairo; text and pagination identical in all three editions]

Bibliography

Fathy, H. (1986) *Natural Energy and Vernacular Architecture: Principles and Examples with Reference to Hot Arid Climates*, Chicago, University of Chicago Press

Fathy, H. (1988) 'Palaces of Mud', *Resurgence*, 103 (March/April 1984), pp. 16–17

Featherstone, A. W. (2002) 'The Principles of Earth Restoration', in Jackson and Svensson (2002) pp. 27–8

Fernandes, E. and Varley, A., eds (1998) *Illegal Cities: Law and Urban Change in Developing Countries*, London, Zed Books

Field, P. (1999) 'The Anti-Roads Movement: The Struggle of Memory Against Forgetting', in Jordan and Lent (1999), pp. 68–79

Fitzgerald, F. (1987) *Cities on a Hill: A Journey Through Contemporary American Cultures*, New York, Touchstone

Foucault, M. (1967) *Madness and Civilization: A History of Insanity in the Age of Reason*, London, Tavistock

Foucault, M. (1979) *Discipline and Punish: The Birth of the Prison*, Harmondsworth, Penguin [first published 1975 as *Surveiller et punir: Naissance de la prison*, Paris, Gallimard]

Frampton, K. (2001) 'Modernization and Local Culture: the Eighth Cycle of the Aga Khan Award', in Aga Khan Award for Architecture (2001) pp. 9–16

Francis, R. (1997) *Transcendental Utopias: Individual and Community at Brook Farm, Fruitlands, and Walden*, Ithaca (NY), Cornell University Press

Freire, P. (1972) *Pedagogy of the Oppressed*, Harmondsworth, Penguin

Fremion, Y. (2002) *Orgasms of History: 3000 years of Spontaneous Insurrection*, Edinburgh, AK Press

Freud, S. (1946) *Civilization and its Discontents*, 3rd edn, London, Hogarth Press [first published 1930, trans. from *Das Unbehagen in der Kultur*, 1929, Vienna]

Gablik, S. (1991) *The Reenchantment of Art*, London, Thames and Hudson

Gablik, S. (1995) *Conversations Before the End of Time*, London, Thames and Hudson

Gandhi, M. K. (1958–1994) *The Collected Works of Mahatma Gandhi*, 100 vols, ed. Swaminathan, K., New Delhi, Ministry of Information and Broadcasting, Government of India

Gandy, M. (2002) *Concrete and Clay: Reworking Nature in New York City*, Cambridge (MA), MIT

Gare, A. E. (1995) *Postmodernism and the Environmental Crisis*, London, Routledge

Generalitat de Catalunya (2001) *Cerdà: The Barcelona Extension (Eixample)*, Barcelona, Institut d'Estudis Territorials, Generalitat de Catalunya

Geoghegan, V. (1984) 'Socialism', in Eccleshal. *et al.* (1984) pp. 91–117

George, S. (2004) *Another World is Possible If . . .*, London, Verso

Geus, M. de (1999) *Ecological Utopias: Envisioning the Sustainable Society*, Utrecht, International Books

Ghirardo, D. (1996) *Architecture after Modernism*, London, Thames and Hudson

Giddens, A. (1990) *The Consequences of Modernity*, Cambridge (UK), Polity Press

Gimeno, E. (2001) 'The Birth of the Barcelona Extension *(Eixample)*', in Generalitat de Catalunya (2001), pp. 20–3

Goldfinger, E. (1993) *Villages in the Sun: Mediterranean Community Architecture*, New York, Rizzoli

Goodwin, B. (1978) *Social Science and Utopia*, Hassocks (Sussex), Harvester Press

Griffin, R., ed. (1995) *Fascism*, Oxford, Oxford University Press

Grosz, E. (2002) 'The Time of Architecture', in Bingaman *et al.* (2002) pp. 265–78

Grosz, E. (2004) *The Nick of Time: Politics, Evolution, and the Untimely*, Durham (NC), Duke University Press

Guérin, D. (1998) *No Gods No Masters: An Anthology of Anarchism* (Book 1), Edinburgh, AK Press

Guha, R. (1989) *The Unquiet Woods: Ecological Change and Peasant Resistance in the Himalaya*, New Delhi, Oxford University Press

Guha, R. and Martinez-Alier, J. (1997) *Varieties of Environmentalism: Essays North and South*, London, Earthscan

Hall, P. and Ward, C. (1998) *Sociable Cities: The Legacy of Ebenezer Howard*, Chichester, Wiley

Hamdi, N. (1995) *Housing Without Houses: Participation, Flexibility, Enablement*, London, Intermediate Technology Publications

Hardy, D. and Ward, C. (2004) *Arcadia for All: The Legacy of a Makeshift Landscape*, Nottingham, Five Leaves [first published 1984, London, Mansell]

Harland, E. (1993) *Eco-renovation: The Ecological Home Improvement Guide*, Totnes (Devon), Green Books

Hasan, A. (1997) 'The Orangi Pilot Project Housing Programme, Karachi, Pakistan', in Serageldin (1997) pp. 82–5

Haverkort, B., van t'Hooft, K. and Hiemstra, W., eds. (2003) *Ancient Roots, New Shoots: Endogenous Development in Practice*, London, Zed Books

Hearn, J. and Roseneil, S., eds (1999) *Consuming Cultures: Power and Resistance*, Basingstoke, Macmillan

Heller, C. (1999) *Ecology and Everyday Life: Rethinking the Desire for Nature*, Montréal, Black Rose Books

Herbert, R. L. (2002) 'Architecture in Léger's Essays', in Blau and Troy (2002) pp. 77–88

Herzen, A. (1979) *From the Other Shore* and *The Russian People and Socialism*, Oxford, Oxford University Press

Herzen, A. (1980) *Childhood, Youth and Exile*, Oxford, Oxford University Press

Hess, T. B. and Ashbery, J., eds (1968) *Avant-Garde Art*, London, Collier-Macmillan

Hill, C. (1975) *The World Turned Upside Down: Radical Ideas During the English Revolution*, Harmondsworth, Penguin

Hiller, S. ed. (1991) *The Myth of Primitivism: Perspectives on Art*, London, Thames and Hudson

Holland, M. (1993) 'Reclaiming Community: One Neighbourhood's Efforts to Stave off Ghetto-ization and Gentrification – With the Help of Cohousing', *In Context*, 35 (Spring), pp. 31–2

Holston, J. ed. (1999) *Cities and Citizenship*, Durham (NC), Duke University Press

Hosagrahar, J. (2005) *Indigenous Modernities: Negotiating Architecture and Urbanism*, London, Routledge

Hostetler, J. A. and Huntington, G. E. (1967) *The Huttites in North America*, New York, Holt, Reinhart and Winston

Houtart, F. and Polet, F., eds (2001) *The Other Davos: the Globalization of Resistance to the World Economic System*, London, Zed Books

Howard, E. (1902) *Garden Cities of Tomorrow*, London, Swan and Sonneschein

Hughes, J. and Sadler, S., eds (2000) *Non-Plan: Essays on Freedom, Participation and Change in Modern Architecture and Urbanism*, Oxford, Architectural Press

Hundert, E. J. (1997) 'The European Enlightenment and the History of the Self', in Porter (1997) pp. 72–83

Huq, R. (1999) 'The Rightto Rave: Opposition to the Criminal Justice and Public Order Act 1994', in Jordan and Lent (1999) pp. 15–34
Idris, M. and Peng, K. K. (1990) 'The Sarawak Natives' Defence of the Forests', in Woodhouse (1990) pp. 165–71
Illich, I. (1971) *De-Schooling Society*, New York, Harper and Row
Illich, I. (1973) *Tools for Conviviality*, London, Calder and Boyars
Illich, I. (1986) *H20 and the Waters of Forgetfulness*, London, Marion Boyars
Illich, I. (1996) *The Right to Useful Unemployment and its Professional Enemies*, London, Marion Boyars [first published 1978, London, Marion Boyars]
Jackson, H. (2002a) 'Food Production: Locally and in Ecovillage Design', in Jackson and Svensson (2002) pp. 31–3
Jackson, H. (2002b) 'Integral Societies: From Cohousing to Ecovillages', in Jackson and Svensson (2002) pp. 156–8
Jackson, H. and Svensson, K. (2002) *Ecovillage Living: Restoring the Earth and Her People*, Dartington (Devon), Green Books
Jacob, J. (1997) *New Pioneers: The Back-to-the-Land Movement and the Search for a Sustainable Future*, University Park (PA), Pennsylvania State University Press
James, L. (1993) 'From Robinson to Robina and Beyond: Robinson Crusoe as a Utopian Concept', in Kumar and Bann (1993) pp. 33–45
Jamison, A. (2001) *The Making of Green Knowledge: Environmental Politics and Cultural Transformation*, Cambridge (UK), Cambridge University Press
Jeffries, R. (1895) *After London, or Wild England*, London, Cassell
Jordan, T. (2002) *Activism: Direct Action, Hacktivism and the Future of Society*, London, Reaktion
Jordan, T. and Lent, A. (1999) *Storming the Millennium: The New Politics of Change*, London, Lawrence & Wishart
Joseph, M. (2002) 'Frugality and the City: Hanoi Palimpsest', in Bingaman *et al.* (2002) pp. 242–55
Jullian, P. (1977) *The Orientalists*, Oxford, Phaidon
Kabeer, N. (1994) *Reversed Realities: Gender, Hierarchies and Development Thought*, London, Verso
Karnouk, L. (1988) *Modern Egyptian Art: The Emergence of a National Style*, Cairo, American University, Cairo Press
Keil, R., Bell, D. V. J., Penz, P. and Fawcett, L., eds (1998) *Political Ecology: Global and Local*, London, Routledge
Kennedy, J. F., Smith, M. G. and Wanek, C., eds (2002) *The Art of Natural Building*, Gabriola Island (BC), New Society Publishers
Klein, N. (1988) *Love, Guilt and Reparation, and Other Works 1921–1945*, London, Virago
Kleiner, D. (2000) ''Case Studies of Eco-Neighbourhoods', in Barton (2000) pp. 268–82
Kleinhenz, C. and LeMoine, F. J. (1999) *Fearful Hope: Approaching the New Millennium*, Madison (WI), University of Wisconsin Press
Kohn, N., (1970) *The Pursuit of the Millennium*, London, Paladin
Kristeva, J. (2002) *Revolt, She Said*, Los Angeles (CA), Semiotext(s)
Kroll, L. (1980) 'Architecture and Bureaucracy', in Mikellides (1980) pp. 162–70
Kropotkin, P. (1902) *Mutual Aid*, London, Heinemann [popular edition 1915, London, Heinemann]
Kropotkin, P. (1919) *Fields, Factories and Workshops*, London, T. Nelson & Sons
Kumar, K. and Bann, S., eds (1993) *Utopias and the Millennium*, London, Reaktion

Kuspit, D. (1993) *The Cult of the Avant-Garde Artist*, Cambridge (UK), Cambridge University Press

Laclau, E. (1996) *Emancipation(s)*, London, Verso

Lacour, C. B. (1996) *Lines of Thought: Discourse, Architectonics, and the Origin of Modern Philosophy*, Durham (NC), Duke University Press

Larkin, R. and Foss, D. (1988) 'Lexicon of Folk Etymology', in Sayres (1988) p. 363

Larner, J. W. Jr (1962) *Nails and Sundrie Medicines: Town Planning and Public Health in the Harmony Society 1805–1840*, Harrisburg (PA), Pennsylvania Historical and Museum Commission [reprinted from *Western Pennsylvania Historical Magazine*, 45(3), pp. 209–27, September 1962]

Lauritsen, P. W. (2002) *A Short Guide to Christiania*, Copenhagen, Aschehoug Dansk Forlag A/S

Leach, N., ed. (1999) *Architecture and Revolution: Contemporary Perspectives on Central and Eastern Europe*, London, Routledge

Leary, T. (1970) *The Politics of Ecstasy*, London, Paladin

Le Corbusier (1946) *Towards a New Architecture*, London, Architectural Press [first published 1923 as *Vers une architecture*, Paris, Editions Crès]

Le Corbusier (1987a) *The City of Tomorrow and Its Planning*, New York, Dover [first published 1929, London, Routledge and New York, Payson and Charles Ltd; translated from the 8th edition of *Urbanisme*, first published in French in 1924]

Le Corbusier (1987b) *Aircraft*, London, Trefoil Press [first published 1935]

Lee, M. F. (1995) *Earth First! Environmental Apocalypse*, Syracuse (NY), Syracuse University Press

Lefebvre, H. (1991) *The Production of Space*, Oxford, Blackwell

Léger, F. (1970) 'Popular Dancing', in Tate Gallery catalogue (1970), *Léger and Purist Paris*, pp. 93–4 [first published 1925 in *Bulletin de L'Effort Moderne*, 12–13, February and March]

Lerner, S. (1998) *Eco-Pioneers: Practical Visionaries Solving Today's Environmental Problems*, Cambridge (MA), MIT

Levi, P. (1987) *If This is a Man*, London, Sphere

Levine, F. (1990) 'Dividing the Water: The Impact of Water Rights Adjudication on New Mexican Communities', *Journal of the Southwest*, 32 (3), pp. 268–77

Light, A. and Smith, J. M., eds (2005) *The Aesthetics of Everyday Life*, New York, Columbia University Press

Lloyd, J. (1991) 'Emil Nolde's Ethnographic Still Lifes: Primitivism, Tradition, and Modernity', in Hiller (1991) pp. 90–112

Lloyd, R. (2006) *Neo-Bohemia: Art and Commerce in the Postindustrial City*, New York, Routledge

Lockwood, G. B. (1905) *The New Harmony Movement*, New York, D. Appleton and Company

Lodziak, C. (2002) *The Myth of Consumerism*, London, Pluto

Lorde, A. (2000) 'The Master's Tools Will Never Dismantle the Master's House', in Rendell *et al.* (2000) pp. 53–5

Low, S. (2000) *On the Plaza: The Politics of Public Space and Culture*, Austin (TX), University of Texas Press

McCreery, S. (2001) 'The Claremont Road Situation', in Borden *et al.* (2001), pp. 229–45

McEvoy, K. (2006) Living in Intentional Community', *Permaculture*, 48, pp. 16–19

McHenry, P. G. Jr (1996) *The Adobe Story: A Global Treasure*, Washington (DC), American Association for International Aging

McKay, G. (1996) *Senseless Acts of Beauty: Cultures of Resistance since the Sixties*, London, Verso

McKeon, M. (1987) *The Origins of the English Novel, 1600–1740*, London, Radius [2nd edn 1988, Baltimore (MD), Johns Hopkins University Press]

McLaughlin, C. and Davidson, G. (1985) *Builders of the Dawn: Community Lifestyles in a Changing World*, Summertown (TN), Book Publishing Company

Maïni, S. (2002) *Auroville Building Centre Earth Unit: Earthen Architecture for Sustainable Habitat*, Auroville, CDRC

Manuel, F. E., ed. (1973) *Utopias and Utopian Thought*, London, Souvenir Press

Marcuse, H. (1956) *Eros and Civilization*, London, Routledge and Kegan Paul

Marcuse, H. (1968a) 'Liberation from the Affluent Society', in Cooper (1968) pp. 175–92

Marcuse, H. (1968b) *Negations*, Harmondsworth, Penguin

Marcuse, H. (1969) *An Essay on Liberation*, Harmondsworth, Penguin

Marcuse, H. (1970) *Five Lectures*, Harmondsworth, Penguin

Marcuse, H. (1998) *Technology, War and Fascism*, Collected Papers vo. 1, ed. Kellner, D., London: Routledge

Marcuse, H. (2007) *Art and Liberation*, Collected Papers vol. 4, ed. Kellner, D., London, Routledge

Markus, T. A. (2002) 'Is There a Built Form for Non-Patriarchal Utopias?', in Bingaman *et al.* (2002) pp. 15–32

Marsh, J. (1982) *Back to the Land: The Pastoral Impulse in Victorian England from 1880 to 1914*, London, Quartet Books

Martineau, A. (1986) *Herbert Marcuse's Utopia*, Montréal, Harvest House

Massey, D. (1994) *Space, Place and Gender*, Cambridge (UK), Polity Press

Mather, F. C. (1965) *Chartism*, London, Historical Association

Max-Neef, M. (1982) *From the Outside Looking In: Experiences in Barefoot Economies*, Uppsala (Sweden), Dag Hammarskjöld Foundation

Maxwell, K. (2002) 'Lisbon: The Earthquake of 1755 and the Urban Recovery under Marquês de Pombal', in Ockman (2002) pp. 20–45

Melehy, H. (1997) *Writing Cogito: Montaigne, Descartes and the Institution of the Modern Subject*, Albany (NY), State University of New York Press

Melish, J. (2003) *Harmony in 1811*, Harmony (PA), Historic Harmony Inc. [first published 1812]

Meller, H. E., ed. (1979) *The Ideal City*, Leicester, Leicester University Press [including Canon Barnett, 'The Ideal City' (1893–4) and Patrick Geddes, 'Civics: As Applied Sociology' (1906)]

Meller, H. (2001) *European Cities 1890–1930s: History, Culture and the Built Environment*, Chichester, Wiley

Meltzer, G. (1999) 'Co-housing: Linking Communitarianism and Sustainability', *Communal Societies: Journal of the Communal Studies Association*, 19 [accessed online: www.aiid.bee.quit.edu.au/~meltzer/articles/articles.ht]

Melville, H. (1972) *Typee*, Harmondsworth, Penguin

Merkel, J. (2003) *Radical Simplicity: Small Footprints on a Finite Earth*, Gabriola Island (BC), New Society Press

Meyer, M. C. (1984) *Water in the Hispanic Southwest: A Social and Legal History*, Tucson (AZ), University of Arizona Press

Midelfort, H. C. E. (1999) 'Madness and the Millennium at Münster, 1534–1535', in Kleinhenz and LeMoine (1999) pp. 115–34

Mikellides, B., ed. (1980) *Architecture for People*, London, Studio Vista

Miles, M. (2004a) *Urban Avant-Gardes: Art, Architecture and Change*, London, Routledge
Miles, M. (2004b) 'Drawn and Quartered: El Raval and the Haussmannization of Barcelona', in Bell and Jayne (2004) pp. 37–55
Miles, M. (2007) *Cities and Cultures*, London, Routledge
Miles, S. and Miles, M. (2004) *Consuming Cities*, Basingstoke, Palgrave
Miller, Daniel (1991) 'Primitive Art and the Necessity of Primitivism to Art', in Hiller (1991) pp. 50–71
Miller, David (1992) 'Community and Citizenship', in Avineri and de-Shalit (1992) pp. 85–100
Miller, T. (2002) 'The Sixties-Era Communes', in Braunstein and Doyle (2002) pp. 327–52
Mollison, B. (1988) *Permaculture: A Designer's Manual*, Tyalgum (NSW), Tagari Publications
Monkerud, D., ed. (2000) *Free Land, Free Love: Tale of a Wilderness Commune*, Aptos (CA), Black Bear Mining and Publishing Company
Moore-Lappé, F. (1990) 'Food First: Beyond Charity, Towards Common Interests', in Woodhouse (1990) pp. 127–38
More, T. (1965) [1516–17] *Utopia*, Harmondsworth, Penguin
More, T. (1997) [1516–17] *Utopia*, New York, Dover
Morley, H. (1887) *Ideal Commonwealths*, 3rd edition, London, George Routledge and Sons
Mumford, L. (1973) 'Utopia, the City and the Machine', in Manuel (1973) pp. 3–24
Murray, I. Z. (2002) 'The Burden of Cubism: the French Imprint on Czech Architecture, 1910–1914', in Blau and Troy (2002) pp. 41–58
Naess, A. (1989) *Ecology, Community and Lifestyle*, Cambridge (UK), Cambridge University Press
Naess, A. (1995) 'The Deep Ecology Movement: Some Philosophical Aspects', in Sessions (1995) pp. 64–84 [reprinted from *Philosophical Inquiry*, 8 (1), 1986]
Nerval, G. de (1972) *Journey to the Orient*, London, Peter Owen
Nochlin, L. (1968) 'The Invention of the Avant-Garde: France, 1830–80', in Hess and Ashbery (1968) pp. 1–24
Noorman, J. K. and Uiterkamp, T. S. (1998) *Green Households? Domestic Consumers, Environment and Sustainability*, London, Earthscan
Norberg-Hodge, H. (2002) 'Community Supported Agriculture', in Jackson and Svensson (2002) p. 39
Nordhoff, C. (1966) *The Communistic Societies of the United States from Personal Observations*, New York, Dover [first published 1875]
Ockman, J., ed. (2002) *Out of Ground Zero: Case Studies in Urban Reinvention*, Munich, Prestel Verlag
Overy, P. (2002) 'The Cell in the City', in Blau and Troy (2002) pp. 117–40
Owen, R. (1972) *A New Vision of Society*, London, Dent, Everyman's Library [first published in Everyman's Library 1927, including texts from 1813 to 1820]
Özkan, S. (1997) 'Architecture to Change the World', in Serageldin (1997) pp. 41–6
Papanek, V. (1984) *Design for the Real World*, London, Thames and Hudson
Papanek, V. (1995) *The Green Imperative: Ecology and Ethics in Design and Architecture*, London, Thames and Hudson
Peet, R. and Watts, M., eds (1996) *Liberation Ecologies: Environment, Development, Social movements*, London, Routledge
Pepper, D. (1991) *Communes and the Green Vision: Counterculture, Lifestyle and the New Age*, London, Green Print (Merlin Press)
Pepper, D. (1996) *Modern Environmentalism: An Introduction*, London, Routledge

Petegorsky, D. W. (1999) *Left-Wing Democracy in the English Civil War*, London, Sandpiper [first published 1940, London, Gollancz]

Pickles, J. (2004) *A History of Spaces: Cartographic Reason, Mapping and the Geo-coded World*, London, Routledge

Pinder, D. (2005) *Visions of the City*, Edinburgh, Edinburgh University Press

Pinto, V. (1998) *Gandhi's Vision and Values: The Moral Quest for Change in Agriculture*, New Delhi, Sage

Plows, A. (1995a) 'Eco-philosophical and Popular Protest: The Significance and Implications of the Ideology and Actions of the Donga Tribe', *Alternative Futures and Popular Protest*, vol. 1 [not paginated]

Plows, A. (1995b) 'The Donga Tribe: Practical Paganism Comes Full Circle', *Creative Mind*, 27, pp. 25–7

Pointon, M., ed. (1994) *Art Apart: Art Institutions and Ideology Across England and North America*, Manchester, Manchester University Press

Porter, R., ed. (1997) *Rewriting the Self: Histories from the Renaissance to the Present*, London, Routledge

Pradervand, P. (1990) *Listening to Africa; Developing Africa from the Grassroots*, New York, Praeger

Rangan, H. (1996) 'From Chipko to Uttaranchal: Development, Environment and Social Protest in the Garwhal Himalays, India', in Peet and Watts (1996) pp. 205–26

Raputov, L. and Lang, M. H. (1991) 'Capital City as Garden City: The Planning of Postwar Revolutionary Moscow', in *The Planning of Capital Cities*, 7th International Planning History Conference, Thessaloniki, pp. 795–812

Reibel, D. B. (2002) *Old Economy Village*, Mechanicsburg (PA), Stackpole Books

Rendell, J., Penner, B. and Borden, I., eds (2000) *Gender Space Architecture: An Interdisciplinary Introduction*, London, Routledge

Richards, G. (1991) *The Philosophy of Gandhi*, Richmond (Surrey), Curzon Press

Richards, J. M., Serageldin, I. and Rastorfer, D. (1985) *Hassan Fathy*, Singapore, Concept Media (and London, Architectural Press)

Robinson, J. (2006) *Ordinary Cities: Between Modernity and Development*, London, Routledge

Rose, G. (1993) *Feminism and Geography: The Limits of Geographical Knowledge*, Cambridge (UK), Polity Press

Ross, K. (1988) *The Emergence of Social Space: Rimbaud and the Paris Commune*, London, Macmillan

Roszak, T. (1995) *The Making of a Counter Culture*, revised edition, Berkeley, University of California Press [first published 1969, New York, Doubleday]

Roy, A. (2001) *Power Politics*, 2nd edition, Cambridge (MA), South End Press

Rugg, J. and Hinchcliffe, D., eds (2001) *Recoveries and Reclamations*, Bristol, Intellect Books

Ryan, M. (1997) *Civic Wars: Democracy and Public Life in the American City during the Nineteenth Century*, Berkeley, University of California Press

Sadler, S. (1999) *The Situationist City*, Cambridge (MA), MIT

Said, E. (1985) *Orientalism: Western Conceptions of the Orient*, Harmondsworth, Penguin [first published 1978, London, Routledge and Kegan Paul]

Sakolsky, R. and Koehnline, J. (1993) *Gone to Croatan: Origins of North American Dropout Culture*, New York, Autonomedia

Salecl, R. (1999) 'The State as a Work of Art: The Trauma of Ceausescu's Disneyland', in Leach (1999) pp. 92–111

Sandercock, L. (1998) *Towards Cosmopolis*, Chichester, Wiley
Sandweis, S. (1998) 'The Social Construction of Environmental Justice', in Camacho (1998) pp. 31–57
Sayres, S. (1988) *The 60s Without Apology*, Minnesota, University of Minneapolis Press
Schwartz, W. and Schwartz, D. (1998) *Living Lightly on the Earth: Travels in Post-Consumer Society*, Charlbury (Oxon), Jon Carpenter
Scrase, T. (1983) 'Planning History', Bristol, Bristol Polytechnic [distance learning material]
Searle, D., ed. (1997) *Gathering Force: DIY Culture – Radical Action for Those Tired of Waiting*, London, The Big Issue Writers
Searle, J. (1972) *The Campus War: A Sympathetic Look at the University in Agony*, Harmondsworth, Penguin
Senghor, L. (1988) *Ce que je crois*, Paris, Grasset
Sennett, R. (1995) *Flesh and Stone: The Body and the City in Western Civilization*, London, Faber and Faber
Serageldin, I., ed. (1997) *The Architecture of Empowerment: People, Shelter and Livable Cities*, London, Academy Editions
Sessions, G., ed. (1995) *Deep Ecology for the 21st Century*, Boston (MA), Shambala
Sheller, M. and Urry, J. (2004) *Tourism Mobilities: Places to Play, Places in Play*, London, Routledge
Shields, R., ed. (1992) *Lifestyle Shopping: The Subject of Consumption*, London, Routledge
Shields, R. (1999) *Lefebvre, Love and Struggle: Spatial Dialectics*, London, Routledge
Shirer, W. L. (1981) *Gandhi: A Memoir*, London, Abacus
Shiva, V. (2000) *Stolen Harvest: The Hijacking of the Global Food Supply*, London, Zed Books
Shklar, J. (1973) 'The Political Theory of Utopia: From Melancholy to Nostalgia', in Manuel (1973) pp. 101–15
Short, J. R. (1996) *The Urban Order: An Introduction to Cities, Culture and Power*, Oxford, Blackwell
Shorthose, J. (2004) 'Nottingham's *de facto* Cultural Quarter: The Lacemarkets, Independents and a Convivial Ecology', in Bell and Jayne (2004) pp. 149–62
Sibley, D. (1995) *Geographies of Exclusion*, London, Routledge
Smith, R. (1997) 'Self-Reflection and the Self', in Porter (1997) pp. 49–57
Snyder, G. (1957) *Earth House Hold*, New York, New Directions Books
Soja, E. (2000) *Postmetropolis: Critical Studies of Cities and Regions*, Oxford, Blackwell
Soleri, P. (1993) *Arcosanti: An Urban Laboratory?*, Scottsdale (AZ), Cosanti Press
Soleri, P. (2001) *The Urban Ideal: Conversations with Paolo Soleri*, Berkeley (CA), Berkeley Hills Books
Soria y Puig, A., ed. (1999) *Cerdà: Five Books of the General Theory of Urbanization*, Madrid, Electa [English edition]
Squires, J. (1994) 'Private Lives, Secluded Places, Privacy and Political Possibility', *Environment and Planning D*, 12, pp. 387–410
Starhawk (2000) *Webs of Power: Notes from the Global Uprising*, Gabriola Island (BC), New Society Publishers
Stead, D. (2002) 'Unsustainable Settlements', in Barton (2000) pp. 29–45
Steegmuller, F., ed. (1983) *Flaubert in Egypt*, London, Michael Haag
Steele, J. (1988) *Hassan Fathy*, Architectural Monograph 13, London, Academy Editions
Steele, J. (1997) *An Architecture for People: The Complete Works of Hassan Fathy*, London, Thames and Hudson

Steele, J. (1998) *Rethinking Modernism for the Developing World: The Complete Architecture of Balkrishna Doshi*, London, Thames and Hudson
Storr, A. (1976) *The Dynamics of Creation*, Harmondsworth, Penguin
Svensson, K. (2002) 'What is an Ecovillage?', in Jackson and Svensson (2002) pp. 10–12
Tafuri, M. (1976) *Architecture and Utopia: Design and Capitalist Development*, Cambridge (MA), MIT
Tarragó, S. (2001) 'Three Holistic Proposals for a Complete, New, Integrated City', in Generalitat de Catalunya (2001) pp. 5–9
Taylor, B. (1994) 'From Penitentiary to "Temple of Art": Early Metaphors of Improvement at the Millbank Tate', in Pointon (1994) pp. 9–32
Taylor, L. (1990) *Housing: Symbol, Structure, Site*, New York, Rizzoli
Taylor, N. (2000) 'Eco-Villages: Dream and Reality', in Barton (2000) pp. 19–28
Thompson, M. (1997) 'Roads and Transport', in Searle (1997) pp. 34–49
Tillyard, E. M. W. (1963) *The Elizabethan World Picture*, Harmondsworth, Penguin
Tocqueville, A. de (1998) *Democracy in America*, London, Wordsworth Press
Toulmin, S. (1990) *Cosmopolis: The Hidden Agenda of Modernity*, Chicago, University of Chicago Press
Traore, M. B. (1990) 'Le Développement, mirroir et écran', Nyon (Switzerland), International Foundation for Development Alternatives, dossier 80
Traugot, M. (1994) *A Short History of the Farm*, Summertown (TN) [published by the author]
Turner, J. F. C. (1963) 'Dwelling Resources in South America', *Architectural Design*, August, pp. 360–93
Turner, J. F. C. (1976) *Housing by People*, London, Marion Boyars
Twinn, C. (2003) 'BedZED', *The Arup Journal*, January, pp. 10–16
Ulam, A. (1973) 'Socialism and Utopia', in Manuel (1973) pp. 116–34
Umenyilora, C. (2000) 'Empowering the Self-Builder', in Hughes and Sadler (2000) pp. 210–21
Vale, B. and Vale, R. (1991) *Green Architecture: Design for a Sustainable Future*, London, Thames and Hudson
Varon, J. (2004) *Bringing the War Home: The Weather Underground, the Red Army Faction, and Revolutionary Violence in the Sixties and Seventies*, Berkeley, University of California Press
Venn, C. (2000) *Occidentalism*, London, Sage
Wagner, P. (2001) *Theorizing Modernity*, London, Sage
Wall, D. (1999) *Earth First! and the Anti-Roads Movement*, London, Routledge
Warburton, D. (1998) *Community and Sustainable Development: Participation in the Future*, London, Earthscan
Ward, C. (1991) *Influences: Voices of Creative Dissent*, Hartland (Devon), Green Books
Warnke, M. (1994) *Political Landscape: The Art History of Nature*, London, Reaktion
Wheeler, J. (2007) 'Tolerance Fades in Denmark as Christiania Free Town Faces New Era', www.worldpress.org/Europe/2647.cfm
White, J. (1972) *The Birth and Rebirth of Pictorial Space*, London, Faber and Faber
Whitefield, P. (1993) *Permaculture in a Nutshell*, East Meon (Hampshire), Permanent Publications
Wigley, M. (1992) 'Untitled: The Housing of Gender', in Colomina (1992) pp. 327–89
Williams, R. (1961) *The Long Revolution*, London, Chatto and Windus
Williams, R. (1977) *Marxism and Literature*, Oxford, Oxford University Press
Williams, R. (1981) *Culture*, London, Fontana

Williams, R. (1989) *The Politics of Modernism*, London, Verso
Wilson, E. (1991) *The Sphinx in the City: Urban Life, the Control of Disorder, and Women*, Berkeley, University of California Press
Wilson, E. (2003) *Bohemians: The Glamorous Outsiders*, 2nd edn, London, I. B. Tauris
Wilson, P. W. (1995) *Pirate Utopias: Moorish Corsairs and European Renegades*, New York, Antonomedia [2nd revised edn 2003]
Winslow Cohousing (1993) 'Winslow Cohousing: A Self-portrait: Members of North AmerIca's First Resident-developed Cohousing Group Reflect on a Year of Living in Community', *In Context*, 35 (Spring), pp. 39–42
Woodhouse, T., ed. (1990) *Replenishing the Earth: The Right Livelihood Awards 1986–1989*, Totnes (Devon), Green Books
Wrench, T. (2000) *Building a Low Impact Roundhouse*, East Meon (Hampshire), Permanent Publications
Yapa, L. (1996) 'Improved Seeds and Constructed Scarcity', in Peet and Watts (1996) pp. 69–85
Zerzan, J. (2002) *Running on Emptiness: The Pathology of Civilization*, Los Angeles, Feral House
Zimmerer, K. S. (1996) 'Discourses on Soil Erosion in Bolivia: Sustainability and the Search for Socio-environmental "Middle Ground"', in Peet and Watts (1996) pp. 110–24
Zukin, S. (1995) *The Cultures of Cities*, Oxford, Blackwell

Index

95th Street Collective (New York), 83

Abode, The, 84
Admiralstrasse (Germany), 74
Adorno, Theodor W., 18, 96
Aga Khan Award for Architecture, 159
Agarwal, Bina, 92–93
Ahmedabad (India), 153
Albert, M., 101
Alberti, Leon Battista, 28
Alcyone Community (Oregon), 121
Alfassa, Mirra, *see* The Mother
Algiers, 59, 67–68, 69, 185
Alphen-aan-den-Rijn (Utrecht), 74
Amaurot, 13
Ambridge, 180, 183
Amin, Samir, 101
Anabaptist Free Commonwealth, 15
Anabaptists, 8
Anarchist Land Colony, 51
Andra Pradesh, 93
Anger, Roger, 191
Anti-university of London, 79
Antliff, M. 69
Antwerp (Belgium), 9, 11, 43
Anvers *see* Antwerp
Anydrus, 13
Applehurst, 52
Aranya (India), 133, 153, 154–155
Arcadia, 7
Arcosanti (Arizona), 3, 121, 181–188, 211
Arendt, Hannah, 4, 111
Ashbury Street, 81
Ashton, O. 51
Assam (India), 169
Aswan (Egypt), 135, 140
Aswan High Dam, 151
Aurobindo, Sri, 191, 194
Auroville (India), 4, 128, 165, 189–194
Auroville Building Centre, 192

Austerlitz, 15
Aztec civilization, 136

Babeuf, François-Noel, 41
Bacon, Francis, 7, 18, 29
Bahuguna, Sunderlal, 93–94
Bainbridge Island (Washington state), 211
Baltimore, 144
Bang, Jan Martin, 102, 120–121, 127, 222
Banham, Reyner, 185
Barcelona, 59–62, 75
Barefoot College (Rajasthan, India), 3, 133, 157–174, 228, 229
Barker, Francis, 25, 26, 101
Barleus, Caspar, 25, 32, 33
Barnett, Canon, 67
Barry, John, 104
Barton, Hugh, 66–67, 74
Bat'a, Tomás, 71
Baudelaire, Charles, 46
Bauman, Zygmunt, 30, 98
Bear Tribe Medicine Society, 89
Beaver County (Pennsylvania), 179
Beck, Garrick, 87
Beddington, 72
BedZED (Beddington Zero Energy Development), 59, 72–74
Beecher, Jonathan, 38–39, 40, 42, 43, 44, 45, 46
Beeckman, Isaac, 26
Beevers, Robert, 64
Bellamy, Edward, 66
Belley (France), 40, 43
Belsey, Catherine, 10
Belzig (Germany), 220
Bender, Thomas, 66
Benjamin, Walter, 23, 44, 70
Berg, Peter, 84, 86
Berger, Morroe, 146
Berg-Schlosser, Dirk, 134

Berlin Wall, 202
Bernardin de St Pierre, Jacques-Henri, 19
Bernstein, Richard, 34
Berry, Kate, 95
Besançon (France), 38, 40
Beuys, Joseph, 79
Beyond Petroleum, 115
Bienvenu, Richard, 38–39, 40, 42, 43, 44, 45, 46
Bihar, 93, 169
Bingaman, Amy, 16
BioRegional Development Group, 72
Bird, John, 206
Birmingham Town Hall, 50
Bjerre, Ria, 198
Black Bear Ranch (California), 84, 89
Black Panthers, 77
Blackman, Winifred, 138
Blackwood, Caroline, 114
Blake, William, 112
Blanchard, Tamsin, 124
Bloch, Ernst, 11, 12–13, 16, 17, 20–21, 36
Boal, Augusto, 96
Boghari, 68
Bolman, Robert, 126
Bonaparte, Napoleon, 50
Bonsdorff, Pauline von, 198–199
Bookchin, Murray, 82
Booth, Charles, 66
Bourneville, 63, 64, 66, 67, 140
Bové, Jose, 96
Boyer, C. 206
Bradford (England), 47
Bread and Roses Collective, 83
Breda (Netherlands), 26
Bretting, John, 96
Bristol, 67
Brithdir Mawr (Wales), 120, 126
Bronstein, Jamie, 52
Bronx, 85
Brook Farm, 47, 175
Broome, Jon, 67
Brown, Norman, 82
Bruno, Giordano, 80
Buber, Martin, 55
Buckingham, James Silk, 63
Buenfil, Alberto, 78, 82, 84, 86, 87–88
Buenos Aires, 62
Burkino-Faso, 92
Byker Wall (Newcastle), 74

Cadbury, 64, 66
Cairo, 134, 135, 138, 139, 140, 146
Cambridge Cohousing (Massachusetts), 4, 123, 209–213
Cambridgeshire, 50
Campanella, Tommasso, 7, 16–18, 37
Cape Frio, 11
Capra, Fritjof, 101
Carey Peter, 11, 16, 21, 35, 41, 55, 66, 105
Carmen, Raff, 95–96, 97
Carney, J. A., 96
Ceausescu, Nicolai, 68
Çelik, Z., 69
Central Park (New York), 66
Centrefor Alternative Living Medicine (CALM), 87
Cerdà, Idelfons, 59–62, 66, 75, 227
Césaire, Aimé, 89, 172
Chamber of Deputies, 54
Chamber of peers, 54
Chambers, Robert, 97, 104
Chapman, Emma, 124, 125
Charterville Allotments, 37, 51–52, 125
Chartist Land Company, 51, 52–54, 66
Chartists, 3, 52–53
Chicago, 63
Chinook Community, 84
Chipko Movement, 93–94
Choay, Françoise, 34
Chomsky, Noam, 82
Christiania (Copenhagen), 4, 195–200, 226
CIAM, *see* International Congress of Modern Architecture
Cilliers, Paul, 106
Cimitière des Innocents, 32
Claeys, Gregory, 41
Claremont Road (east London), 113, 114, 115, 117, 119
Clark, Kim, 211
Clark, W. 95
Clousden Hill, 51
Coates, Chris, 8, 15, 35, 46, 47, 49, 50, 51, 52, 53, 66
Cobb, John Jr., 184
Cold War, 76, 81, 174,
Colomina, Beatriz, 69
Concordium, 46
Connor, Steven, 32
Considérant, Victor, 47
Cook, Jeffrey, 184–185
Cook, Thomas, 135
Coombes, Annie, 104
Corbin, 33
Cornell University (Ithaca), 82, 215, 217
Cosgrove, Denis, 28
Cottingham, John, 26

250 Index

Country Joe, 81
Coventry Peace House, 157, 175–176, 229
Crouch, David, 67, 123
Crystal Palace, 64
Crystal Waters (Australia), 126–128
Curtis, B., 151
Curtis, Kimberly, 70, 111

Damisch, Hubert, 28
Darwin, Charles, 106, 231
Davidson, Gordon, 83–84, 85, 89–90, 121
Davis (California), 211
Dean, Andrea O., 129
Defoe, Daniel, 7, 19
Degen, M., 59
Delhi, 157, 171
Deloria, Philip, 89
Desai, Pooran, 105
Descartes, René, 3, 22–23, 24, 25–31, 37, 227
Dethier, Jean, 134, 148
Deutsche, Ludwig, 135
Dewey, John, 29
Didion, Joan, 75–76, 78, 81
Diggers and Dreamers, 86
Diggers, 8, 15, 35–36, 46, 53, 54, 83, 112
District, The, 75, 77, 78–79, 80, 82
Dobson, Andrew, 113, 114–115
Docker, John, 89
Dominguez, Joe, 104
Donald, James, 52, 68
Dongas, 110, 115, 118–119, 223
Doshi, Balkrishna, 133, 152–154
Doughty, D., 134
Douglas Firs, 124
Doxiades Associates, 146
Doyle, Michael, 76, 81, 82–83,
Dra Valley (Morocco), 148
Dregger, Leila, 224
Dufour, 96
Dunster, Bill, 72–74
Dürrschmidt, Jörg, 98
Duss, John S., 183
Dutton, Rachel, 106
Dwyer, Bill, 112

Earth First! (EF), 111, 115–116, 118
East Devon, 117
East Side Service Organisation, The (ESSO), 82
East Wahdat (Jordan), 153
Ecker, Achim, 129–130, 222
Ecolonia, 74

Economy Village (Pennsylvania), 179–183
Ecovillage (Ithaca), 4, 214–219
Eden, 7, 19, 41, 62
Edwards, Amelia, 135
Edwards, Stewart, 55
Egyptian Society of Architects, 147
Eixample, 59–60
Ekram, Lailun, 154
Elliott, Jennifer, 94, 105
Elyot, Sir Thomas, 14
Engels, Friedrich, 38, 52, 62
Erasmus of Rotterdam, 12
Erskine, Ralph, 74
Escobar, Arturo, 95, 97, 104, 105
Ezbet Al-Basry, 138

Fairfield, 84
Fairlie, Simon, 123–124, 125, 126
Fanon, Frantz, 89, 136, 152, 172
Farber, David, 81
Farm, The, 84, 113
Farrell, James J., 46, 80, 81, 83
Farren, Mick, 112
Fathy, Hassan, 3, 133–152, 155–156, 157, 159, 171, 228
Fawzi-Bey Kalinia, 138
Feast Hall, 179
Featherstone, Alan Watson, 122
Fernandes, Edesio, 134
Fever Hospital (India), 157–159, 165, 167
Field, Patrick, 113, 114, 119
Findhorm Community (Scotland), 86, 122, 212
First Garden City Limited, 65
Fish, The, 81
Fitzgerald, Frances, 80
Flaubert, Gustave, 135
Foss, Daniel, 89
Foucault, Michel, 32, 54–55
Fourier, François Marie Charles, 3, 37–48, 49, 50, 51, 54–55, 56, 59, 77, 80, 102, 108, 127, 181, 227
Frampton, Kenneth, 159–160, 173
Francis, Richard, 39, 40, 45, 46, 47, 55, 102, 175
Free University (Berlin), 79
Freee Art Collective, 231, 232
Freire, Paolo, 96, 173–174
Fremion, Yves, 8
French Revolution, 37
Freud, Sigmund, 56
Friedman, Mitch, 116
Friends of Art and Life, 138
Futurists, 82

Gablik, Suzi, 88, 106
Galileo, Galilei, 31
Gandhi Labour Institute, 153
Gandhi, Mahatma, 92, 93, 159, 160, 162–164, 172, 191
Gandy, Matthew, 66
Garden City Association, *see* First Garden City Limited
Garden City, 59, 62–67, 71, 75, 128, 227
Garden City's Central Council, 66
Garden of Eden, *see* Eden
Gare, Arren, 95, 97
Garhwal, 93
Gaskin, Stephen, 84, 85
Gauguin, Paul, 20
Geoghegan, Vincent, 49
George, Henry, 66
George, Susan, 99
Gerome, Jean-Leon, 135
Geus, Marius de, 56
Ghirardo, Diana, 74, 210, 211
Giddens, Anthony, 73
Gilles, Peter, 12
Gilson, Etienne, 27
Gimeno, 60
Glasgow, 144
Glen Affric, 122
Gloucestershire, 52
Godwin, Francis, 20, 21, 56
Golden Gate Park (San Francisco), 78, 81
Golders Green, 63
Goldfinger, Myron, 134
Goldhaft, Judy, 84, 86
Goodman, Barry, 128
Goodman, Paul, 82
Goodwin, Barbara, 45, 55
Gourna al-Jadida, 133, 134, 140, 142
Grand Avenue, 64
Grand Canyon Uranium Mine, 116
Grateful Dead, 81
Gray, Eileen, 69
Greenham Common, 114
Greenham Women's Movement, 118
Greenpeace, 107, 118
Griffin, Roger, 30
Grosz, Elizabeth, 12, 14, 15, 106, 231
Guérin, Daniel, 51
Guevara, Ernesto, 81
Guha, Ramachandra, 92, 93, 96

Haight Street, 81
Haight-Ashbury (San Francisco), 75, 78, 80; *see also* The District
Hall, Peter, 63, 65, 66
Halle (Germany), 43
Hamdi, Nabeel, 97, 134, 153,155
Hampshire, 46, 115
Hanseatic League, 23
Hansom, Joseph, 50
Haranya, 138
Hardy, Dennis, 52, 67, 125
Harland, Edward, 105
Harmony (Pennsylvania),49, 180–181
Harmony Hall (Hampshire), 46, 50
Harmony Society, 180
Harvey, William, 28
Hasan, A., 153
Haussmann, Baron, 23, 62, 68
Haverkort, Bertus, 105
Heller, Chaia, 107–108, 112, 116
Henderson, Ethel, 65
Henry VIII, 8, 11
Heronsgate, 52, 53
Hertfordshire, 52, 53
Herzen, Alexander, 56
Hill, Christopher, 8, 9, 35, 36
Hilversum Meent (Netherlands), 210
Himalayas, 164
Hippies, 76, 78, 80, 86–87
Hobbes, Thoms, 24
Hodson, William, 50
Hoffman, Allen, 82
Hog Farmers (California), 78, 82, 86
Holland, Matt, 210
Holocaust, 30
Hooker, Richard, 14
Hooper, Daniel, *see* Swampy
Horkheimer, Max, 18, 97
Hornsey School of Art (London), 79
Hosagrahar, Jyoti, 134, 172
Hostetler, John Andrew, 15
Housing Cooperative Movement, 71
Houtart, François, 96, 100, 101
Howard, Ebenezer, 59, 62, 63, 64–67, 71, 75
Hoxton (London), 207
Hundert, E. J., 24–25
Huntington, G. E., 15
Huq, R., 111
Hursley, Timothy, 129
Hyde Park Diggers, 113
Hythloday, 9, 11, 13, 15, 34

Idris, M., 92
Illich, Ivan, 32–33, 76, 206
Institute for the Formation of Character, 48
International Congress of Modern Architecture (CIAM), 70, 151

Jackson, Hildur, 120, 122–123
Jacob, Jeffrey, 125, 126
James, Louis, 11, 16
Jamison, Andrew, 91, 107, 108, 126
Jeanneret, Pierre, 69
Jefferson Airplane, 79
Jeffries, Richard, 105
Jerome, Judson, 86
John of Leyden, 15
Jordan, Tim, 101, 111, 112, 117
Joseph, May, 174–175
Jullian, Philippe, 136

Kabeer, Naila, 96
Karachi, 153
Karlsruhe (Germany), 28
Karnataka (India), 92
Karnouk, Liliane, 136
Kate Richards O'Hare Collective, 83
Keil, Roger, 96
Kennedy, Joseph F., 105, 129, 134
Kerala, 93
Kett, Robert, 15
Keynes, Milton, 64
Kharga, 146
Khartoum, 135
Klein, Naomi, 32, 69
Kleiner, Deborah, 74, 216
Kohn, Norman, 8, 35
Kristeva, Julia, 79–80, 226
Kroll, Lucien, 74
Kropotkin, Peter, 51, 66, 127
Kuspit, Donald, 88
Kyd, Thomas, 10

L'Ange, François-Joseph, 39
L'Enfant, Charles, 28
La Haye (France), 26
Laclau, Ernesto, 228
Lacour, Claudia Brodsky, 22, 27, 29, 31
Ladakh (Afghanistan), 164
Laforgue, Jules, 76
Lake Brienne, 19
Lama Foundation, 83
Lancaster, Joseph, 48
Lang, Fritz, 201
Lang, M. H., 71
Larkin, Ralph, 89
Larner, John W. Jr., 180
Lauritsen, Pernille, 197, 198, 199
Le Corbusier (Charles-Édouard Jeanneret), 3, 59, 67–72, 75, 185, 191, 227
Leary, Timothy, 80
Lebensgarten, 86, 122

Lee, Martha F., 115, 116
Leeds, 64
Lefebvre, Henri, 4, 70–71, 77, 111, 116, 117, 119, 152, 231
Léger, Fernand, 70
Leipzig (Germany), 43
Lerner, Steve, 125
Letchworth, 65, 67
Levellers, 8, 35, 36
Lever, 64
Levi, Primo, 30
Levine, F., 95
Lewis, John Frederick, 135
Lindegger, Max, 128
Lisbon, 23
Lloyd, Richard, 88, 207
Lockwood, George, 47, 180, 181
Lodziak, Conrad, 88
Lollards, 8
London Underground, 116
London, 23, 51, 52, 62, 63, 64, 72, 86, 103, 113, 115, 144
Long, Leo de, 210
Lorde, A., 129
Low, Setha, 17, 29
Lowbands, 52
Lu'Luat al-Sahara, 146
Ludvigsen, Jacob, 195
Luisenstadt (Germany), 74
Luther, Martin, 9
Luxor, 135, 143
Lyon (France), 38, 39, 40

McCreery, Sandy, 114, 119
McEvoy, K., 120
McHenry, 185
McKay, George, 111, 112–113, 118–119
McKeon, Michael, 16
McLaughlin, Corrine, 83–84, 85, 89–90, 121
Madhya Pradesh, 169
Madison Gardens (New York), 183
Madrid, 60
Maïni, Satprem, 192, 194
Malmö (Sweden), 210
Manchester airport, 117
Manchester, 52
Mandeville, Bernard, 24–25
Manea Fen, 50
Manuel, Frank, 41, 45, 46
Marcuse, Herbert, 4, 20, 37, 46, 56, 77, 79, 82, 229–231
Markus, Thomas, 15
Marlowe, Christopher, 10

Marrakesh, 148
Marsh, Jan, 113
Martineau, Alain, 90
Martinez-Alier, Juan, 92, 96
Marx, Karl, 38, 54, 77, 96, 119
Marxist-Maoists, 87
Massey, Doreen, 28
Mather, F. C., 51
Matrimandir, 189, 191, 192
Max-Neef, Manfred, 95
Maxwell, D., 23
Mazdoor Kisan Shakti Sangathan
   (Labour Farmer Strength Organisation),
   167
Melehy, Hassan, 27, 29
Melia, 138
Melish, John, 180
Meller, Helen, 66, 67, 71, 83
Meltzer, Graham, 210, 211
Melville, Herman, 7, 19
Mercury, 17
Merkel, Jim, 103
Merry Pranksters, 78, 82, 86
Mesa City, 185
Meyer, Michael C., 95
Midelfort, H. C. E., 8
Miles, Malcolm, 59, 69, 99, 119, 231
Miles, Steven, 62
Miller, David, 85, 88
Miller, Timothy, 83, 84, 85
Minster Lovell (Oxfordshire), 52
Mithras, 14
MKSS, *see* Mazdoor Kisan Shakti
   Sangathan
Mockbee, Samuel, 129
Mollison, Bill, 120
Monkerud, Don, 84
Montaigne, Michel, 26, 37
Moore-Lappé, Frances, 104
Mor, 17
More, Thomas, 1, 7, 8–16, 29, 32, 34, 37,
   39, 46, 51, 54
Morea, Ben, 82
Morea, Janice, 82
Morelly, Abbé, 41
Morley, Henry, 9
Morton, John, 12
Mother, The, 191, 192
Motherfuckers, 78, 82, 84, 86
Mount Athos, 12
Mudhead Kachinas, 116
Müller, Bernhard, 181
Mumford, Lewis, 31
Münster, 15

Müntzer, Thomas, 8
Murnau, F. W., 201

Nadu, Tamil, 164
Naess, Arne, 101–102, 104, 118
Naples, 16
Narmada Valley (India), 92
National Community Friendly Society, 50
Nelson, Elizabeth, 112
Neruda, Pablo, 226
Nerval, Gérard de, 135
Neville, John, 7
New Baris, 133, 146–149, 153
New Bucharest, 68
New Buffalo (New Mexico), 84
New Gourna, 133, 134–135, 137,
   138–146, 147, 150, 151, 171
New Harmony (Indiana), 47, 181
New Labour, 54
New Lanark (Scotland), 46, 48–49, 50, 66
New Life, 83
New Orleans, 180
New School for Social Research, 80
New Spain, 29
New York, 4, 66, 144
New York Diggers, 82–83,
New York Stock Exchange (NYSE), 82
Newcastle, 51, 73
Nile, 133, 135, 137
Nile Valley, 146
Nochlin, Linda, 55
Noorman, Klaas Jan, 105
Norberg-Hodge, Helena, 123
Nordhoff, Charles, 181, 183
Norfolk, 15
Nottingham, 206
Nubia, 135, 140

Obus Plan, 69, 70, 185
O'Connor, Feargus, 51–52, 53
Oldcastle, Sir Thomas, 8
Olds, Rob, 106
Orangi Pilot Project Housing Programme,
   153
Orissa, 169
Orleans (France), 43
Osiris, 17
Ottoman Empire, 135
Overy, P., 69
Owen, Robert, 3, 37, 46–49, 50, 51, 54,
   66, 180
Owen, Robert Dale, 181
Oxleas Wood (east London), 113
Özkan, Suha, 150

254 Index

Pacific Northwest, 87
Papanek, Victor, 88, 105, 174
Paris, 23, 28, 33, 35–36, 40, 43, 44, 54, 59, 62, 64, 68, 69, 75, 76, 83, 116, 137, 226
Paris Commune, 76
Parker, Barry, 65, 67
Parliament Hill Fields, 79
Peace Camp (Greenham Common), 114
Peet, Richard, 96
Peng, K. K., 92
Pepper, David, 36, 51, 99, 106, 113, 115
Pestalozzi, Heinrich, 48
Petegorsky, David, 8, 15
Philadelphia, 180
Phillip II, 17
Phun City, 112
Pickles, John, 18, 28
Pinder, David, 76–77, 83
Pinto, Vivek, 163–164
Pisanello, 13
Place Dauphine, 28
Place Royale, 28
Planet Drum, 86
Plato, 14, 29
Plows, Alex, 118–119
Polet, François, 96, 101
Pollock, Jackson, 88
Polylerits, 12
Polynesia, 19
Pondicherry, 191
Port Sunlight, 63, 64, 140
Porto Allegre (Brazil), 96, 101
Potsdam (Germany), 28
Pradervand, Pierre, 92, 96
Prindeville, Diane-Michelle, 96
Proudhon, Pierre-Joseph, 55
Providence Zen Centre, 84
Puritans, 107

Quakers, 35, 48
Queensland (Australia), 127

Rainbow Family, 87
Rainbow Gatherings, 87, 110, 119, 223
Rajasthan (India), 3, 164, 165, 169, 228
Rangan, Haripriya, 93–94, 96, 102
Ranters, 35
Rapp, Frederick, 179, 180–181
Rapp, George, 179, 180
Rappite community, 47, 49
Raputov, L. 71
Rastorfer, Darl, 140, 148, 150, 151
Reclaim the Streets (RTS), 115, 116–117
Regent Street, 23

Reibel, Daniel B., 180, 181
Rembrandt, Harmenszoon van Rein, 25
Richards, Glyn, 162–163
Richards, J. M., 134, 142, 147, 150, 151–152
Richardson, Brian, 67
Richelieu, Cardinal, 17
Riddlestone, Sue, 105
Rimbaud, Arthur, 76
Rio de Janeiro, 68
Robin, Vicki, 104
Robinson, Jennifer, 69
Rome, 12
Rose, Gillian, 28
Ross, Kristin, 76–77
Roszak, Theodore, 79
Rothko, Mark, 88
Roundhouse Trust, 120
Rousseau, Jean-Jacques, 7, 18–19, 40
Roy, Arundhati, 92
Roy, Bunker, 157
Royal Palace, 54
Rue de Seine (Saint-Germain-des-Prés), 83
Rural Studio, 129
Ryan, Mary, 66

Sadler, Simon, 47, 76
Saettedammen (Denmark), 120
Said, Edward, 136, 137
Said, Hamed, 134
St George's Hill (England), 15, 35, 36, 112
Saint-Simon, Henri de, 55
Salecl, Renata, 68–69
Salt, Titus, 47
Saltaire (England), 47
San Fernando Valley (California), 86
San Francisco, 75–77, 84, 85, 87, 112, 119, 227
San Francisco Diggers, 75, 77, 78, 81–82, 84, 86, 111
San Francisco Mime Troupe, 75, 78, 80–81
Sandercock, Leonie, 34, 97, 107
Sanders, Lise, 16
Sandweis, Stephen, 95
Santa Fe, 86, 185, 186
Sarawak, 92
Sartre, Jean-Paul, 89
Savvaerket (Denmark), 210
Schwartz, Dorothy, 85, 124, 127, 128
Schwartz, Walter, 85, 124, 127, 128
Scottsdale (Arizona), 186
Scrase, Tony, 66
Searle, Denise, 79

Seattle, Chief, 104
Seddon, Thomas, 135
Segal, Walter, 67
Sellier, Henri, 71
Senghor, Léopold, 89, 172
Sennett, Richard, 28, 63
Serageldin, Ismaïl, 134, 153, 156
Seventh Heaven Jazz Café, 114
Shakespeare, William, 10
Shields, Rob, 70, 77, 111, 116–117, 231
Shirer, William, 162
Shiva, Vandana, 92
Shklar, J., 16
Short, John Rennie, 23
Shorthose, Jim, 206
Sibley, David, 32–33
Sinai, 146
Situationism, 75, 76, 77, 83, 116, 119
Smith, Roger, 25
Snigs End, 52
Snyder, Gary, 87–88
Social Work Research Centre (SWRC), *see* Barefoot College
Soja, Edward, 23
Solborg Camphill Village (Norway), 120
Soleri, Paolo, 184–186
Somerset, 73
Soria y Puig, Arturo, 60, 61, 62
Sousa, John Philip, 183
South Seas, 7
Starhawk, 111–112
Stead, Dominic, 198
Steegmuller, Francis,135
Steele, James, 134, 138, 140, 146, 148, 151, 152–154
Stonehenge, 112
Storr, Anthony, 25
Strawberry Lake, 87
Students for a Democratic Society (SDS), 77, 78
Sub-Saharan Africa, 92
Suez Canal, 135
Summer of Love, 75, 76, 77–78, 83, 112, 119, 133, 227
Summertown (Tennessee), 84
Suresnes (France), 71
Svanhom (Denmark), 120
Svensson, Karen, 120, 121–122
Swampy, 117
Szerszynski, Bronislaw, 91, 107, 108–109, 117, 126

Tafuri, Manfredo, 70
Tahiti, 20, 43

Talkha (Egypt), 135
Tamil Nadu (India), 4, 189, 192
Taos (New Mexico), 134
Tap, Robert, 128
Tarragó, S., 61
Tate, Henry, 62
Taylor, B., 62
Taylor, Lisa, 68
Taylor, Nigel, 73
Tehri-Garhwal, 93
Tenochtitlan (Mexico), 17
Theban necropolis, 133
Thompson, Mick, 113–114
Thoreau, Henry David, 102–103
Tillyard, E. M. W., 14
Tilonia (Rajasthan), 133, 164, 171
Tinker's Bubble (Somerset), 73, 123, 124–125, 175
Tinngården (Denmark), 210
Tinngården 2 (Denmark), 210
Tipi Valley (Wales), 113, 118
Tir Ysbrydol, 120
Toulmin, Stephen, 24, 31–32, 34
Trafalgar Square, 23
Trancho, Jane, 211
Trans Europe Halles Network, 202
Traore, M. B., 95
Traugot, Michael, 85
Tribal Council, 87
Turner, John, 134, 159
Turner, Paul, 12
Twinn, C., 72
Twyford Down, 110, 115, 117–118, 119

Ufa-Fabrik (Berlin), 4, 201–204
Uiterkamp, Ton Schoot, 105
Ulam, Adam, 56
Ulm (Germany), 26
Umenyilora, Chinedu, 153, 155
Union of Contemporary Architects, 71
Unwin, Raymond, 65, 67
Uttaranchal, 94, 169
Uzupio (Lithuania), 4, 205–208

Vale, Bredna, 67
Vale, Robert, 67
Valois, George, 69
Vandkunsten, Tegnestuevo, 210
Varley, Ann, 134
Varon, Jeremy, 76, 77, 89, 116
Vastu-Shilpa Foundation for Studies and Research in Environmental Design, 154
Venn, Couze, 29, 89, 90
Verona, 13

256 Index

Versailles, 28
Vespucci, Amerigo, 11
Vibild, Karsten, 210
Vietnam War, 76, 77, 79
Village Industries Association, 163
village of Economy (Pennsylvania), the, 3
Vilnius, 207
Voisin Plan, 69

Wagner, Peter, 24, 29, 34
Wakefield, Edward Gibbon, 63
Walden Pond, 102, 103
Waldhör, Ivo, 210
Walker, Liz, 216
Wall, Derek, 111, 113, 117
Ward, Colin, 52, 56, 63, 64, 65, 66, 67, 125
Warnke, Martin, 104
Washington DC, 64
Wassif, Ramsis Wissa, 138
Watts, Michael, 96
Weathermen, 115
Weeda, Pieter, 210
Wheeler, J., 199
White, John, 28
Whitefield, Patrick, 120
Whitwell, Stedman, 47, 180
Wigley, M., 28
Williams, Raymond, 10, 98, 107, 109, 206–207, 228

Wilson, Elizabeth, 65, 67, 207
Windsor Free Festivals, 112
Windsor Great Park, 112
Winslow Cohousing, 211
Winstanley, Gerrard, 35
Wittenberg (Germany), 9
Woodstock, 112
World Social Forum, 96, 100
Wrench, Tony, 120
Wright, Frank Lloyd, 185
Württemberg, 180

Yamuna, 93
Yapa, L., 92
Yippies, 77, 82
Young, Geoff, 128
Yugoslavia, 24

ZEGG, *see* Zentrum für Experimentelle GesellschaftsGestaltung
Zentrum für Experimentelle GesellschaftsGestaltung (Centre for Experimental Culture Design) (Germany), 4, 46, 121, 122, 127, 129, 220–226, 228, 229
Zerzan, John, 99
Zimmerer, K. S., 92
Zlín, 71–72, 75
Zola, Emile, 44
Zorach, Rebecca, 16
Zukin, Sharon, 206, 208

eBooks – at www.eBookstore.tandf.co.uk

A library at your fingertips!

eBooks are electronic versions of printed books. You can store them on your PC/laptop or browse them online.

They have advantages for anyone needing rapid access to a wide variety of published, copyright information.

eBooks can help your research by enabling you to bookmark chapters, annotate text and use instant searches to find specific words or phrases. Several eBook files would fit on even a small laptop or PDA.

NEW: Save money by eSubscribing: cheap, online access to any eBook for as long as you need it.

Annual subscription packages

We now offer special low-cost bulk subscriptions to packages of eBooks in certain subject areas. These are available to libraries or to individuals.

For more information please contact webmaster.ebooks@tandf.co.uk

We're continually developing the eBook concept, so keep up to date by visiting the website.

www.eBookstore.tandf.co.uk

Routledge Critical Introductions to Urbanism and the City

Series edited by **Malcolm Miles**, University of Plymouth, UK and **John Rennie-Short**, University of Maryland, USA

The series is designed to allow undergraduate readers to make sense of, and find a critical way into, urbanism.

* covers a broad range of themes
* introduces key ideas and sources
* introduces complex arguments clearly and accessibly
* are affordable and well designed
* reconsiders cities in relation to a specific theme
* allows the authors to articulate their own position
* bridges disciplines, and theory and practice

The series will cover social, political, economic, cultural and spatial concerns. It will appeal to students in architecture, cultural studies, geography, popular culture, sociology, urban studies, urban planning. While being firmly situated in the present, it also introduces material from the cities of modernity and post-modernity which has fed into that position. Each volume will approach cities in a trans-disciplinary way.

Out Now:

Cities and Cultures
Malcolm Miles, University of Plymouth, UK

Cities and Consumption
Mark Jayne, Manchester University, UK

Cities and Economies
Yeong-Hyun Kim, Ohio University, USA and John Rennie-Short, University of Maryland, USA

Cities and Nature
Lisa Benton-Short, George Washington University, USA and John Rennie-Short, University of Maryland, USA

Forthcoming to this series:

Cities and Cinema
Barbara Mennel, University of Florida, USA

Digital Cities
Chris Benner, Pennsylvania State University, USA

Urban Erotics
David Bell and Jon Binnie. Both at Manchester Metropolitan University, UK

www.routledge.com/geography